John Howard Redfield, William Morris Davis, Edward Lothrop Rand

Flora of Mount Desert Island Maine

a preliminary catalogue of the plants growing on Mount Desert and the adjacent

islands

John Howard Redfield, William Morris Davis, Edward Lothrop Rand

Flora of Mount Desert Island Maine
a preliminary catalogue of the plants growing on Mount Desert and the adjacent islands

ISBN/EAN: 9783744718653

Printed in Europe, USA, Canada, Australia, Japan

Cover: Foto ©berggeist007 / pixelio.de

More available books at **www.hansebooks.com**

FLORA

OF

MOUNT DESERT ISLAND, MAINE.

Flora of Mount Desert Island, Maine.

A

PRELIMINARY CATALOGUE

OF THE

PLANTS GROWING ON MOUNT DESERT

AND THE ADJACENT ISLANDS.

BY

EDWARD L. RAND AND JOHN H. REDFIELD.

With a Geological Introduction

By WILLIAM MORRIS DAVIS,

AND A NEW MAP OF MOUNT DESERT ISLAND.

CAMBRIDGE:

JOHN WILSON AND SON.

University Press.

1894.

CONTENTS.

———•———

	Page
PREFACE	7

GENERAL OUTLINE OF PLAN OF CATALOGUE.

I. Indigenous Plants	13
II. Introduced Plants	13
III. Synonyms	14
IV. Arrangement and Nomenclature	14
V. Citation of Authors	15
VI. Forms	15
VII. Terms denoting Relative Occurrence	16
VIII. Plants not represented in the Herbarium	16
IX. Abbreviations	17
X. Geographical Nomenclature	17

INTRODUCTION.

I. Mount Desert and its Flora	19
II. The Map of Mount Desert Island	28
List of Corrections	31
III. Botanical Nomenclature of the Catalogue	32

OUTLINE OF THE GEOLOGY OF MOUNT DESERT.

Introduction	43
The Granite Belt	46
The Pre-Granitic Rocks	51
The Post-Granitic Rocks	55
The Great Denudation	56
The Glacial Invasion	63
Postglacial History	67

FLORA: CATALOGUE OF PLANTS.

PHANEROGAMIA, OR FLOWERING PLANTS.
DICOTYLEDONES, OR EXOGENOUS PLANTS 75
ANGIOSPERMEÆ : POLYPETALÆ 75
GAMOPETALÆ 107
APETALÆ 139
GYMNOSPERMEÆ 149
MONOCOTYLEDONES, OR ENDOGENOUS PLANTS 150

CRYPTOGAMIA, OR FLOWERLESS PLANTS.
PTERIDOPHYTA 184
BRYOPHYTA 190
MUSCI 190
HEPATICÆ 219
THALLOPHYTA 227
CHARACEÆ 227
ALGÆ 227
LICHENES 250

SUMMARY 275

APPENDIX. — LIST OF EXCLUDED SPECIES 277

INDEX 281

PREFACE.

THE territory covered by this Catalogue of Plants comprises the Island of Mount Desert and the adjoining islands, the more important of which are the Cranberry Isles, Bartlett Island, Thompson Island, and the Porcupine Islands. The Duck Islands, lying some miles seaward southerly from the Cranberry Isles, are also included for convenience, although having no close connection geographically with the rest of the territory. Politically it comprises the towns of Eden, Mount Desert, Tremont, Cranberry Isles, a small part of Trenton, and a part of Long Island Plantation, in which the Duck Islands are included. All of this territory, with the exception of the Duck Islands, is shown on the map that has been prepared to accompany this Catalogue.

In 1880 the Champlain Society, an association of college students formed for the purpose of field work and study in various branches of natural science, established its camp on the shores of Somes Sound at Wasgatt Cove, Mount Desert Island. This Catalogue of Plants represents the final results of work begun by its botanical department, while the introductory article on the Geology of Mount Desert represents the work of its geological department. Two years later one of the authors, John H. Redfield, began independent investigation of the Island flora. In 1888 the Champlain Society allowed its botanical work to pass into the hands of the other author, Edward L. Rand, who, however, had been connected with the work from its beginning. Soon afterwards the authors consolidated the results of all the botanical work on the Island, so far as they were able, and henceforth carried on the work together, with such assistance as could be procured from other botanists. Although more or less incomplete, and somewhat hastily prepared, this Catalogue is now presented, at the request of many interested in the subject, as a preliminary contribution to a Flora of Mount Desert Island. This is done with the hope that it may serve as a means of exciting interest in the undertaking, and thus make possible a more complete catalogue in the near future.

So far as the study of its flora is concerned, Mount Desert has no history. We are told by the early explorers that wild roses and beach peas were abundant, and that is all. No botanists native to the Island — if any there were or are — have given us information as to its plants. All such information has come from such botanists as have chanced to go there from a distance, usually during the summer months only. Even of these the known list is not long, and only few antedate the beginning of systematic work in 1880. It has, furthermore, been extremely difficult to discover the names of these botanists, and to consult their notes and collections, although the authors have endeavored in many ways to accomplish this. The result naturally has been far from satisfactory. In spite of all these discouragements, however, the work on the Flora has been carried on with perseverance. It is now hoped that from the very fact of the publication of present results help may be obtained for the future that otherwise would have been locked up in the herbaria and note-books of unknown workers in the same territory.

Specimens of every plant in this list, with very few exceptions, will be found preserved in the Mount Desert Herbarium, at present kept in Cambridge, Mass. These exceptions, most of which are either Algæ or Lichens, are denoted by an asterisk prefixed to the name of the species. For specimens of plants thus marked, as well as for other plants from collectors now unrepresented, we shall be most grateful.[1] The Philadelphia Academy of Natural Sciences, furthermore, has an almost complete duplicate set of the Phanerogams and Pteridophyta; and Dr. Carl Warnstorf of Neuruppin, Germany, has a duplicate set of the Sphagna. Duplicates from the Herbarium have also been distributed among various public and private herbaria of the country.

Much care has been taken to make the Catalogue reliable. Very few plants have been admitted to the list except on the authority of an undoubted specimen, and in every case of exception only on a positive affirmation by a specialist or other botanist of high repute as to the authenticity and identity of the lost specimen. Moreover, we have had the kind assistance of many of the leading botanists of the country in the determination of specimens in difficult families and genera, and in cases of doubtful determination, as well as in the criticism and correction of our manuscript. Prof. L. H. Bailey has given his help in Carex and Rubus; Mr. M. S. Bebb, in Salix; Prof. William Trelease, in Rumex and Epilobium; Dr. Thomas C. Porter, in Solidago, Aster, and Mentha;

[1] Any correspondence relating to the Flora may be addressed to Edward L. Rand, 740 Exchange Building, Boston, Mass., or to John H. Redfield, 216 West Logan Square, Philadelphia, Penn.

Mr. John K. Small, in Polygonum; Prof. F. Lamson Scribner, in Gramineæ; Dr. L. M. Underwood, in Isoetes and in Hepaticæ; Mr. George E. Davenport, in Filices; Dr. T. F. Allen, in Characeæ; Messrs. Frank S. Collins and Isaac Holden, in Algæ; Dr. Carl Warnstorf, Prof. D. C. Eaton, and Mr. Edwin Faxon, in Sphagnum; Mrs. E. G. Britton and Dr. Charles R. Barnes, in the other Mosses; Dr. J. W. Eckfeldt, Miss Mary L. Wilson, and Miss Clara E. Cummings, in Lichenes; and Dr. B. L. Robinson, Dr. N. L. Britton, Dr. Thomas Morong, Mr. Walter Deane, and Mr. M. L. Fernald, in various other determinations. The article on the Geology of Mount Desert has been kindly contributed by Prof. William M. Davis, of Harvard College. To these and to all others who have done so much to add to the value and accuracy of this Catalogue, to the various collectors whose names appear therein, and to President Charles W. Eliot of Harvard University, through whose interest and kindness the publication of our work has been made possible, we extend our sincere thanks.

Acknowledging, as we have at the outset, the incompleteness of this Catalogue in many of its divisions, we issue it at the present time to assist those interested in the plants of the Island to the acquirement of a better knowledge of its flora. With this end in view, therefore, it has seemed well to include, for the benefit of specialists, even manifestly incomplete lists of some of the Cryptogams. The list of Vascular Cryptogams (Pteridophyta), the Ferns and their allies, is fairly complete; the lists of Mosses and Liverworts are well advancing towards completion, and the same is true of the lists of Lichens and of the marine Algæ. Very little work, however, has been done thus far in the collection and determination of the fresh-water Algæ and the Fungi, and it has seemed better for the present to omit the latter altogether from this Catalogue. It is hoped that in the near future more attention may be given to increasing in a marked degree our knowledge of this part of the Island flora.

July 1st, 1894.

GENERAL OUTLINE OF THE PLAN OF THE
CATALOGUE OF PLANTS.

I. THE names of plants supposed to be indigenous to North America are printed in heavy broad-face type. "Indigenous" is but a relative term, and can hardly be employed with any accuracy, even in its commonly accepted sense, in connection with the flora of a comparatively small territory, without a very definite knowledge of the facts of local plant introduction and distribution. A list of Mount Desert plants pretending to show the plants "indigenous" to the Island would contain mere guesswork in many cases, and would only lead to much confusion. It has therefore been thought better to draw the distinction between plants indigenous to the continent and those evidently foreign to it, and to add such notes as may seem of value relating to the introduction within our territorial limits, on the one hand, of North American species, and, on the other, of species from other continents.

II. The names of plants believed to be introduced into North America are printed in small capitals. It is to be understood that in nearly every case such plants have been, so far as known, indirectly introduced through other parts of this continent into Mount Desert Island. There is very little of that evidence of direct introduction of any of these plants which is so common about seaports where there is direct communication with foreign countries. This class of plants includes both those that are fully naturalized and those which as yet are only adventive or well established garden escapes. For reasons already given, it will be seen that there are no ballast plants to be catalogued.

III. Synonyms are printed in Italics.

IV. The principle underlying the arrangement and nomenclature of the Catalogue is a very simple one, more practical
than theoretical. It is this: to follow in these respects some
manual or other work of high authority, regardless of any
fancy or preference of the authors. It seems hardly necessary
to state to any one of practical experience, that the office of a
local Flora, or of any similar work designed fully as much for
the public generally as for scientists, is not to serve as a nomen-
clator, or to present an opportunity for the author to display his
fads to his own satisfaction and the confusion of the reader,
but rather to be a help and an aid to a better knowledge
of the plants of any given region. Unless descriptions are
added, so that such a catalogue is in reality a manual in itself,
reference must be made to some well known work or handbook.
Such being the case, the authors have felt obliged to adopt
some such standard as a guide and basis for the arrangement
and nomenclature of the Catalogue,[1] giving only such synonyms
as in their judgment may serve some useful purpose of identification or of information, and making such corrections only as
do not interfere with the system of the guide adopted.

In nomenclature and arrangement, the sixth (revised) edition
of Gray's Manual by Watson and Coulter is followed for the
Phanerogams (Flowering Plants); for the Pteridophyta (Vascular Cryptogams), and for the Hepaticæ. Dr. Carl Warnstorf's
articles on the North American Sphagna, in Vol. XV. (1890)
of the Botanical Gazette, are mainly followed for Sphagnum;
while Lesquereux and James's "Mosses of North America" is
followed for the remainder of the Mosses. Tuckerman's works
are followed as far as possible for the Lichens, and Farlow's
"Marine Algæ of New England," with some marked changes

[1] As, however, the subject of botanical nomenclature has been given
undue prominence of late by some of our American botanists among others,
it has seemed better to the authors to discuss this subject at more length in
the Introduction. Had this not been done, it might be asked why the
rules of the so called Rochester and Madison Codes were not followed
as a standard, — an intentional omission for which there is more than ample
justification.

in classification, arrangement, and nomenclature, for the marine Algæ.[1] In every case where descriptions of genera or species found at Mount Desert do not appear for any reason in these works, the authors have tried to give them in the Catalogue, hoping thereby to render unnecessary any reference to works or articles not readily accessible. It is believed that our plan has been adopted throughout with some slight exceptions, most of which need no explanation.

V. It has, however, been thought well to adopt throughout the Catalogue the parenthetical citation of the original author of the specific or varietal name, a method already long adopted by cryptogamic botanists. Thus *Coptis trifolia*, the common Goldthread, was described in 1753 by Linnæus under the name of *Helleborus trifolius*. In 1798, Salisbury considered that the plant showed well marked generic differences, and assigned it to a new genus, *Coptis*. Our plant therefore bears the binomial, *Coptis trifolia* (L.), Salisb. It must bo borne in mind, however, that the author cited in parentheses is cited only for the specific or the varietal name in the binominal, as the case may be, and is connected with that alone, and not with the binominal itself. To the binominal, the name of the author not cited in parenthesis alone applies. If these distinctions be remembered, many of the objections that have been so forcibly urged against this method of citation seem to lose their weight.

VI. The term "form" — *forma* — has been used for the sake of convenience to indicate slight physiological or structural variations seeming of hardly enough importance to mark a good variety, much less a species, and yet worthy of some notice, perhaps of future study. Allowance once rightly made for variation in nature, it becomes a very complex and difficult matter to decide what is a species, what a variety, what a form, what a variation. Without discussion of the subject, it may be said that it has seemed best to recognize as forms substantially the same variations that are indicated by Dr. N. L.

---

[1] See introductory note to the list of Algæ for a fuller statement of the plan adopted.

Britton in his article "On the Naming of 'Forms' in the New Jersey Catalogue."[1] It is not thought well, however, to attach the name of any author to these so called forms, as the line between a form and a mere variation is generally too shadowy to call for the exercise of any judgment worthy of recognition in the decision that one variation or another should be dignified by the term *forma*. Forms should bear names for the sake of convenience, and if properly named; that name should be preserved to avoid confusion, if the form proves after investigation to be a well defined variety. In such a case, however, to cite the author of the name in parentheses seems to savor more of affectation than of common sense or utility, and if so, why mention the author of a form at all? A line must be drawn somewhere to check the increasing tendency to self-glorification that can at present be so easily gratified on the part of the amateur as well as of the professional botanist. In this Catalogue, therefore, as has been stated, no authors are cited for the names of mere forms. Should, however, any one desire to know them, a goodly number may be found in Dr. Britton's article just referred to.

VII. The usual terms *common, uncommon, frequent, infrequent, occasional, rare,* etc., are used to denote the relative occurrence of the different plants. It must be remembered, however, that these terms apply only to a plant in its proper habitat. Because a seashore plant is "common," no one should expect to find it on the mountains! Where few stations are given for any plant, it does not necessarily follow that it does not occur elsewhere. New stations are likely to be reported at any time for nearly all such plants. It merely indicates, therefore, that thus far collectors have not been successful in detecting any very general distribution of the plant in question.

VIII. An asterisk prefixed to names of plants indicates that a specimen of the plant in question is not to be found in the Mount Desert Herbarium at the present time, although the occurrence of the plant within the territorial limits is undoubted.

[1] See Bull. Torr. Bot. Club, xvii. 121.

IX. The abbreviations used for names of authors will be found in Gray's Manual, Gray's Structural Botany, or in Britton's Catalogue of New Jersey Plants. Other abbreviations either require no explanation, or may be found in any manual or text-book. The abbreviation "R. & R.," as may be readily supposed, refers to the authors of this Catalogue.

X. The geographical nomenclature follows that of the map of Mount Desert Island published in June, 1893, to accompany this work. Its nomenclature is based on certain universal and well recognized laws of nomenclature, among the chief of which are the regard for priority, for firmly established custom, for good taste, and for avoidance of unnecessary confusion. The rules adopted by the U. S. Board on Geographic Names have been followed as far as possible as to form, iu order to secure conformity with the Coast Survey Charts and other government publications. Of the changes caused by these rules, the only one that is likely to be commonly noticed is the avoidance of the possessive whenever this can be done without destroying the euphony of the name or changing the descriptive application. In applying this rule, therefore, the possessive *s* has been retained only where it appeared to be necessary for euphony or to avoid misunderstanding, usually where the name is a Christian name, and sometimes where for special reasons or on account of peculiar usage it seemed impracticable to do otherwise. In all cases, however, where the possessive *s* is retained, the possessive apostrophe has been dropped, since the word should no longer be considered as possessive in sense, but as a word in itself.

2

INTRODUCTION.

I. Mount Desert and its Flora.

MOUNT DESERT ISLAND, called by the Indians Pemetic, lies about one hundred and ten miles east of Portland, on the coast of Maine, and less than half that distance from Rockland on the western shore of Penobscot Bay. Its coast is washed by the Atlantic Ocean on the south, by Blue Hill Bay and its tributaries on the west, and by Frenchman Bay and its tributaries on the east and north. On the northwest Mt. Desert Narrows, a shallow strait connecting the waters of these two bays, is crossed by means of two bridges, connecting Thompson Island with the mainland on the north, and with Mt. Desert Island on the south. The area of the Island may be estimated at about one hundred square miles; its greatest length being about fifteen miles, from Hadley Point in Eden on the north to Bass Harbor Head in Tremont on the south; its greatest breadth, about twelve miles, from Great Head in Eden on the east to The Cape in Tremont on the west. The coast line, especially of the southern and western shores, is extremely irregular. Up the centre of the Island for fully half its length from north to south, through the mountain range, passes the fiord of Somes Sound (or "The River"), a deep arm of the sea, dividing the Island into two almost equal sections. Across the

centre from Western Mt. on the west to Newport Mt. on
the east stretches the granitic range of mountains that
has given Mt. Desert its name, rising almost from the
sea to heights varying from about three hundred to over
fifteen hundred feet. Towards the north the ground slopes
to the farming lands of Eden and the great meadow of
the Northeast Creek, and towards the southwest to the
meadows of Marsh Creek, to Great Heath and the boggy
wilderness below the Hio. Between the peaks of the
granitic range lie deep valleys, filled either by an arm of
the sea, as Somes Sound, or by a lake or pond of more or
less magnitude. These are mountain ponds for the
most part, many of them of great depth, with rocky
shores broken by stretches of sand or gravel beaches.
None of the streams are of much size, and the regularity
even of their natural flow has been greatly diminished by
the wanton destruction of the woods about their water
sheds.[1]

All of these facts, however, are much better explained
by the map itself, and by the article on the geology of
the Island by Professor Davis, kindly contributed by
him for this very purpose. It is better, therefore, in this
place to make no more than the most general statements
in regard to the topography. Neither is it well to attempt
any detailed description of the flora in its relations to
these physical and geological characteristics, for as yet
the evidence seems too fragmentary and disconnected to
prove facts of much value. A few brief statements of a
very general nature, illustrated by a few examples, may
however be of interest to the botanist.

One of the most marked characteristics of the Isl-
and flora is its not only strongly northern, but arctic

[1] See "The Woods of Mt. Desert Island," Garden and Forest, II. 483.

character.[1]   On its coast, enveloped in cold fogs and washed by waters chilled by the arctic currrent, it is no wonder that arctic plants like *Montia fontana* and *Stellaria humifusa* should find a congenial home. Moreover, this character of the flora is shown by the fact that, with one exception, *Lycopodium Selago*, the mountain plants descend to the sea level. Neither on the one hand is the altitude of the mountain summits sufficient to develop an alpine flora, nor on the other hand is the warmth and general character of the lowlands sufficient to bring many of the plants of the middle temperate region thus far up the coast of Maine. The flora, then, may be said to be essentially Canadian, having close relations with the very similar flora of New Brunswick. It also shows, apart from its maritime character, many points of resemblance to the general flora of the White Mountain region.   It is in its special problems, however, that plant distribution becomes of great interest at Mt. Desert, and it may be well, therefore, to consider a few cases by way of illustration.

The return of vegetable life after the glacial period must have taken place along somewhat more contracted lines than are shown to-day. Mt. Desert was then, as now, isolated from the mainland, but was without doubt in a state of greater submergence. It is therefore natural that there should exist in abundance on the mainland many plants that are not found at all on the Island, or are found there only very rarely. The water on the north of the Island is not of great extent or depth at present, yet it appears that some plants, especially those with seeds not easily transported by ordinary means, have

[1] About two hundred and thirty of the flowering plants of Mt. Desert are common to the arctic flora.

always found difficulty in crossing it from the mainland. This difficulty has had its effect in decreasing the Island flora.

Again, in the development of their flora the Cranberry Isles have shown some peculiarities. These islands, once doubtless a part of Mt. Desert, and through it connected with the mainland, were later submerged, and then elevated again to develop their flora independently of Mt. Desert, except so far as the flora of the smaller area came from that of the greater, then doubtless more advanced in the renewal of its vegetation, owing to its greater altitude and consequent earlier elevation. That there was some independent development is well shown by the fact that between the Cranberry Isles and the adjacent portion of Mt. Desert about the Sea Wall there exist some remarkable differences in the flora, as well as some strong points of union. Under almost precisely the same conditions, we find *Corema* near the Sea Wall, but not on the Cranberry Isles; we find *Montia*, *Stellaria humifusa*, and *Rubus Chamæmorus* on the Cranberry Isles, but not on Mt. Desert; we find *Symplocarpus fœtidus* and *Hippuris vulgaris* on the Cranberry Isles and also on Mt. Desert, but at the Sea Wall alone. Such evidence as this may point to the introduction of certain plants on Mt. Desert by way of the Cranberry Isles, while on the other hand doubtless most of the plants of the Cranberry Isles came from Mt. Desert.

It is certainly far from improbable that the more northern plants came to the Cranberry Isles by sea, either from the north in later times, or from the south when these islands first appeared above the sea at the conclusion of the glacial period. If from the north, there would be little opportunity for colonization on the rocky eastern and southeastern coasts of Mt. Desert, — an opportunity, however, which would readily be presented on the low shores, and in the coast marshes and lagoons of the Cranberry Isles. Yet it would seem improbable that all these plants reached the Cranberry Isles only. *Montia* has been found also on the Duck Islands, and might likewise, and as readily, be carried by ocean currents farther on, at least to the westward adjoining shores of Mt. Desert. There in the southwestern part of the Island similar conditions existed for the colonization of these plants as on the Cranberry Isles, yet in fact they do not appear, so far as known. If, on the other hand, they came from the south, remaining behind in the progress of plant life northward after the glacial period, and finding here favorable surroundings for their existence, all the more we should expect to find these plants also in the southwestern projection of Mt. Desert Island. Here the land would be reached earlier in the northward march, and would be found to present the same conditions of soil and of general physical character as the Cranberry Isles. As, however, none of these peculiar plants except *Symplocarpus* and *Hippuris* appear even on this part of Mt. Desert, the evidence at present seems in favor of a later migration from the north, rather than of the much earlier introduction from the south. The whole subject is one of great interest, and will repay careful study.

Another interesting feature of the Mt. Desert flora is shown by the comparatively small representation of introduced foreign plants, especially of weeds of cultivated ground. Excluding garden escapes and a few plants naturalized by intentional introduction, we find that the number of weeds is very small in comparison with that of similar areas in New England. The reason is a very simple one, — the slight development of the Island for agricultural purposes, — an explanation that is fully sustained by the facts.

In earlier times very little attention was paid to farming, doubtless because the physical character of the Island is not of a nature favoring agriculture except under limited or somewhat expensive conditions. The surface is mostly mountainous or rocky, the soil is usually thin and poor, and has often disappeared as a covering, — a result of reckless wood cutting and of the consequent forest fires. Taken as a whole, the north of the Island contains the best farming land; the south, for the most part, is too near the dominant granitic range to furnish deep soil or level ground save under exceptional conditions. Moreover, under these unfavorable conditions there was nothing to encourage farming as a means of support, for there was no market for garden products. It is not strange, therefore, that fishing, lumbering, shipbuilding, and other pursuits, were the more profitable employments of the early settlers. All agricultural operations were conducted on a very limited scale, and for the most part involved nothing more than the cultivation of small vegetable patches for home purposes. These patches were seldom well cared for, and were rarely cultivated in the same spot for more than a year at a time. Of late years, however, it has been found profitable by many landowners to raise vegetables to supply the summer demand at Bar Harbor and the other summer resorts of the Island. Consequently there has been more systematic cultivation of the ground both for agricultural and for horticultural purposes.

In the earlier days of the settlement of the Island, therefore, we should expect to find few of those weeds that constantly need the aid of man to secure and maintain a foothold. Such is the case. The weeds of those days were obliged to adapt themselves to the most hostile conditions. If they could not do this, they lingered on year by year wherever they could maintain a foothold, and then almost disappeared from the flora of the Island. Consequently these weeds were largely of native origin, and not many in number.

But within a few years a new state of things has arisen. Not only have the old weeds been gaining a stronger and stronger foothold, but additions to the list are reported every year, chiefly at Bar Harbor or in its neighborhood, whence they spread to other parts of the Island. Only a few years ago such common weeds as *Portulaca oleracea, Amarantus retroflexus* and *A. albus, Medicago lupulina, Lepidium Virginicum, Mollugo verticillata,* and *Plantago lanceolata* were either unknown or so rare that it was difficult even to secure specimens of them. They are now becoming more and more common, and appearing slowly but surely throughout Mt. Desert. Some of these obtain their foothold through cultivation of the soil, and all seem to come, as many people do, because it is the fashion, taking advantage of the increased means of introduction afforded by the importation of foreign seed, of foreign soil with other plants, of hay, and of the various other methods by which weeds travel about from place to place.

This explanation, it is hoped, will show why so many of the common weeds find no place in this Catalogue. It also shows that at any time such additions to the flora are likely to be reported by any botanist who happens to examine the waste and cultivated grounds and the way-sides of the constantly growing villages and settlements. Of these newcomers it will be well to ascertain and note carefully the date of introduction.

It is interesting also to notice what does not appear, as well as what does appear in this Catalogue. It was once said, indeed, that the flora of Mt. Desert was more remarkable for what it did not include than for what it did, — a statement that our present knowledge of the flora hardly seems to justify Yet there are many important gaps in the Catalogue that it is hard to account for in any satisfactory manner. It can only be said that for some reason or other these missing plants do not occur on this part of the coast, or, in cases where they do occur on the adjacent mainland, that they never were able to cross the water to Mt. Desert Island. It is certain that the latitude is not the cause, for these plants are found much farther north. Doubtless the cold east winds and the sea fogs may have driven back many plants trying to effect a lodgment here; but in that case there should be a marked difference between the flora of the exposed southern and eastern coasts, and that of the northwestern, central, and northern parts of the Island. A study of the Catalogue will show that there is some such difference, but not so marked, we think, that it can be relied on as evidence to any very great extent. It proves, however, that no one can be well acquainted with the flora until he has studied carefully the plants of the country lying north of the main granitic belt, as well as those of the better known and more frequented parts of the Island.

It may be interesting to mention some cases of these missing plants. The Pulse Family, *Leguminosœ*, will furnish a striking instance. The Catalogue shows that the Island flora contains only eighteen species, representing eight genera, obviously a very insufficient representation when we consider that shown by many points farther north with otherwise much the same flora. Of these species, ten arc naturalized on this continent from Europe; two are introduced from other parts of North America; two more, as appears from circumstantial evidence, may also have been so introduced; leaving only four species that are indigenous in the common sense. This would seem to prove that at Mt. Desert there was some obstacle besides climate which leguminous plants found it difficult to surmount. That it is not some hostile condition at the present time appears from the fact that when northern species of this family are introduced on the Island they flourish as well there as elsewhere.

Further instances are the genera Asclepias and Gentiana, and many others, — of which no representative whatever is found, — and a number of species belonging to different genera, which are found northward on the mainland, but not on Mt. Desert. It may be that some day many of these missing plants will reach the Island, but at present their absence seems as unmistakable as it is unaccountable.

For its disappointments, however, the flora makes ample compensation. For so limited and circumscribed an area our territory possesses many plants interesting to any lover of our New England flora, and has contributed some forms that are of interest to the general botanist as well. Even its most common flowers take new and unexpected deepness of color from the cool sea air, and are a constant delight both to botanist and mere flower-lover. Our work has been a labor of love, the fruit of happy days, and the source of pleasant memories. If this Catalogue proves a help to those for whom it is intended, and enables them to share the pleasure we have gained on this wonderful island of Mt. Desert, we shall be more than satisfied.

II. The Map of Mount Desert Island.

Some years ago it became very evident that there was
to be great difficulty in properly indicating stations for
the various Island plants needing such limitation. While
it was necessary in some cases to make the station somewhat indefinite in description in order to guard against
extermination on the one hand by the flower-puller and
the plant-digger, and on the other by the over-zealous
botanist, yet it was necessary in all cases to give a name
to the station that should be both accurate and well
known as a matter of geographical nomenclature. To
some it may seem that this involved merely a reference
to any map of the Island to ascertain the necessary
information, but this was a solution of only a portion of
the difficulty. In the first place the two maps most
readily consulted, the Land Map of Colby and Stuart and
the Coast Survey Map, pay very little attention to the
names of the points of minor interest on the Island. As
such points are often of the greatest botanical interest,
and must be referred to, it was clear that the present
maps would not be of much assistance in these cases. In
the second place, the geographical nomenclature employed
on the Coast Survey Map, and followed in some degree
on the Land Map, is often, we regret to say, absolutely
erroneous. In many a case, indeed, there is no explanation whatever to account for the blunders, except that the
officers in charge of the work must have coined names for
their own use, regardless or in ignorance of the fact that
there might be names already attached to the places in
question. In other cases, by some curious mistake, names
have been carelessly transposed and interchanged. The

natural result of all these errors was to establish two sets of names, one known to those acquainted only with the maps, the other to those who either lived on the Island, or knew the Island independently of map knowledge. Furthermore, the matter of nomenclature was much complicated by the insufferable tendency of summer visitors to give new names, often showing the worst possible taste, to any natural feature that might happen to attract their attention. Such names deserve preservation only in rare cases, and should not be tolerated for a moment unless by lapse of time or by custom the new name has fairly superseded the old for all practical purposes.

To remedy these evils, and to secure a standard for citation in our Catalogue, it was decided to make as thorough an investigation of the geographical nomenclature of the Island as possible, to adopt a system of correct nomenclature, and finally to prepare a map that should set forth the results of our work. For over three years this investigation was carried on, until, in June, 1893, the map was published. If we may judge by what we have heard ourselves, or by what has been reported to us by others, very little fault is found with the nomenclature adopted. It is to be borne in mind that where the nomenclature of our map differs from that of the Coast Survey it is to be explained on one of two grounds: either because the Coast Survey attached a name to the wrong locality, or because it coined a name or substituted one of no authority to replace a name well known and in common use upon the Island. A very striking instance of the error last mentioned is found in the unauthorized use of Turtle Lake for Bubble Pond, or for the oldest name of all, now obsolete, Southeast Pond.

The preparation of this map made necessary much correspondence and much careful investigation of ancient
maps, plans, and records. Such an undertaking could
never have been brought to a successful conclusion had
it not been for the kindly interest shown and the invalu-
able assistance given by natives of the Island who knew
and loved it well. Among these helpers, many of whom we
regret to say we hardly know by name, but whose assistance,
by whatever means it reached us, we value highly, we wish
to give our especial thanks to the Rev. Oliver H. Fernald;
to Mr. Eben M. Hamor, of Eden; to Messrs. T. S. Somes,
George A. Somes, Thomas Bartlett, and A. C. Savage, of
Mt. Desert; to Messrs. W. W. A. Heath and C. M. Holden, of Tremont; and to Mr. P. C. Stover, of Cranberry
Isles; all of whom by inquiry, by personal investigation,
and by advice and criticism have done so much to give
the map its accuracy and merit. To Mr. Fernald, born
and brought up on the Island, and still retaining in his
residence in another part of the State his love for his
native place and his interest in its affairs, we owe the in-
spiration of this undertaking, and to his encouragement
and assistance its final accomplishment.

By the kind permission of Prof. Thomas C. Mendenhall,
Superintendent of the United States Coast Survey, we
have used the Coast Survey Map for the important physical
features, making here and there a few corrections, and
supplying a few omissions. We wish here to express
our appreciation of his courtesy, which has enabled us
to give a much better map to the public. New roads,
the town boundaries, and additional wood roads and paths
have been added, the different post-offices indicated, and
such points of interest named as it seemed would make
a map not only suitable for our purpose, but of value

to any one interested in the Island. The general rules followed in regard to nomenclature have already been explained in the Preface.[1]

It could not have been expected that our map would be either complete or entirely free from error. Since its publication, therefore, effort has been made to discover omissions and mistakes, in the hope that some time in the future we can make any corrections that may be found necessary. We wish at present to call attention to the following list of the more important errors and omissions thus far discovered.

(1) The town boundary between Mt. Desert and Tremont in the territory lying between Somes Sound and Great Pond should begin on the eastern shore of Great Pond at the point shown on the map, and should run in a straight line in a southeasterly direction to a point on the shore of Valley Cove nearly opposite the word "Eagle" on the map. This shows the true boundary some distance to the north of the boundary shown on the map.

(2) At the Quarries on the western shore of Somes Sound a post-office should be added, "Halls Quarry P. O."

(3) The name Western Hio, north of Bass Harbor, it seems, should be applied to the southern end of Norwood Ridge. Where the name now stands on the map, Burnt Mt. should be substituted.

(4) Black Point on Great Cranberry Isle may have to be changed to Flaggs Point.

(5) The small brook at Bar Harbor, flowing into the

---

[1] It may be interesting to note the use of the word "heath" on the Island. It is used to denote a large unwooded bog or swamp, usually a sphagnum bog, very wet, and exceedingly difficult to cross. Many of these heaths contain small ponds or spring holes, and in the wetter parts are floating bogs more or less dangerous and treacherous to any one venturing upon them.

cove opposite the southwestern end of Bar Island, is known as Eddys Brook.

(6) The point next northwest of Cape Levi is known as Parker Point.

(7) The brook rising east of Town Hill, and flowing into the South Branch of Northeast Creek is known as Aunt Betsys Brook.

(8) The brook flowing westward into Clark Cove is known as Meadow Brook.

(9) The marsh by the Salt Pond on Thomas Bay is known as Jones Marsh.

(10) Denning Walk at the Quarries on the western shore of Somes Sound lies farther to the eastward, between the road and the shore, and the position of the name should be changed accordingly.

(11) The course of Sunken Heath Brook is shown incorrectly between Sunken Heath and the road. It should flow directly south from the Heath to the road, not as shown on the map. The remainder of its course, however, is correctly shown.

(12) The name Saul Cliff, on the shore south of Bar Harbor, should be Sols Cliff.

We should be very glad to receive from any one other information that will serve to make the map more perfect and useful to those interested in the Island.

III. Nomenclature of the Catalogue.

As was stated in the Preface, it seems well to the authors to state more at length the reasons why they think it advisable to follow as a standard for this catalogue the nomenclature of the sixth edition of Gray's Manual (so far as it covers the ground), rather than that

of the so called Rochester and Madison Codes, adopted
in 1892 and 1893 by the Botanical Club of the American
Association for the Advancement of Science.

At the outset it should again be stated that we believe
the nomenclature and arrangement of a local Flora should
follow that of some well known or authoritative work or
system. If, on the one hand, the system followed is not
well known, the catalogue will not be of much use to many
for whom it was intended; if, on the other hand, the system does not emanate from some respected authority, it is
folly to attempt to force it on any intelligent person.
The main question for an author or compiler to consider
clearly seems to be, What standard can be followed that
will be most intelligible and most useful to those for
whom it is intended, — to the plant lover of slight botanical knowledge, as well as to the professional botanist of
thorough training? In making the decision the author
need not necessarily follow his own personal inclinations,
— in fact it is not right or expedient for him to do so if
clearness and usefulness must be sacrificed thereby, — his
duty is to help, not to hinder his public, and to yield his
personal preference for the good of others. If he wishes
to express his personal opinions and convictions he can
do this at pleasure through many appropriate channels;
he may speak thus whenever he will. The public cares
very little for the personal convictions and peculiar theories of its servants if unintelligible and practically
useless. It does demand, however, that the servant
should do his duty, and serve the good of the master, not
any private or selfish purpose of his own. It is hard to
see how the contrary can be maintained.

It being granted, therefore, that we must not necessarily
set out our own personal opinions, but must make our

catalogue useful and intelligible, in duty to all who consult it, we had to consider what well recognized standard
we could follow. The choice appeared to lie between
Gray's Manual, mainly the work of our greatest botanist,
and the principles now embodied in their strictest form in
the Rochester Code and extended in the Madison Code.
The chief of these principles one of us had studied for
years, and the other had put to practical use, as a test of
their real value. Moreover, we both felt that the priority
of the specific name should on sound analogies be maintained, in opposition to the well known rule of Dr. Gray
that the first specific name in the right genus should
prevail. Nevertheless, as a result of our deliberation we
have decided that a local Flora at this time without question must follow Gray's Manual, whether or not its authors
agree entirely with the nomenclature of that work; that
to follow strictly the system dictated by the Rochester
Code is utterly impracticable and unwise, for it is neither
consistent in theory nor sound in practice. This conclusion has been reached after long judicial consideration of
the arguments for and against the system of the Rochester
Code, whether practical or theoretical in nature, and with
an earnest desire to approve any really beneficial altera-
tions in the commonly accepted system of botanical nomenclature. We regret, therefore, that the Code, as a whole,
must be condemned for the evil that is in it, and that the
good it contains cannot be utilized in its present form.
As it stands, it seems the work of botanists whose vision
is bounded by the book-shelves of the library and by the
herbarium walls rather than of botanists possessing that
added knowledge and grasp of affairs that is so indis-
pensable to a correct solution of difficulties in such a
practical matter as that of botanical nomenclature.

The mental attitude of the supporters of the Rochester Code seems at first somewhat difficult to explain. If we abandon for a theory of our own well known and established principles sanctioned by the greatest authorities and the soundest analogies, we must justify our action. We have not yet seen any such justification of this Code. It seems that the explanation must lie in the fact that its supporters cannot appreciate that they have a case to prove, and that the burden of proof rests on them alone. If they act in contravention of fundamental principles and of the authority and consensus of the greatest botanists, they must prove to the satisfaction of an intelligent man that they are acting rightly. Even granting that the Code is proved of utility, the rule still applies to every change they seek to make. In fact, however, they assume the contrary, and are open to the gravest criticism for constantly leaning in favor of change; and of blindly following what is apparently their guiding principle, — *Quieta movere.* Where doubt exists, the old and accepted name or identification should be preferred in every case to the new and unproved. We know no reason why botanists should be exempt from following such fundamental rules. If the Code permits the contrary practice, as its advocates take for granted, it cannot be followed.

Thus it appears to be most necessary for these botanists to prove that their system secures advantages that the old system does not possess. If, on the one hand, they claim that it is more sound in theory, it may be said that practical relief, not theoretical relief, is needed. Moreover, their theory is inconsistent within itself, being founded partly on absolute dedication of a name to the public, and partly on the absolute inability of the public to do what it will with its own. Thus we are not only told that an author cannot change a name once published, because it has passed from his control; but we are also no less gravely told that a name once published can never be changed by the public either by usage or in any other manner, — an inconsistency that it is hard to explain in any reasonable manner. If, on the other hand, they claim that in practice strict adherence to priority does away with the uncertainties of individual judgment, and secures absolute certainty in nomenclature for past, present, and future, this assertion may be fairly denied, at least so far as the past is concerned. Any one who has followed the many differences in judgment, and the disagreements as to actual priority, can easily realize that it is a matter requiring much acute and long continued investigation to fix absolutely the historical priority and identity of names. This fact should have deterred many botanists from rushing into print with their new-old names, like children eager to display a new toy, only to discover later that they had been too hasty, and had merely added to the ever increasing host of synonyms. Furthermore, how can it be known that this system will be permanent? Its advocates claim that they not only can violate other theories, and coin artificial rules to secure any desired result, but can as readily disregard and reject many principles of a fundamental nature that are well recognized by practical men of affairs, whether scientists or laymen, and have been so recognized and approved by the greatest botanists. If these can be set aside by any one with a theory of his own, what security have we that the Rochester Code, with all its inconsistencies and objectionable features, will not be set aside in a year or two in favor of some radically different theory? This is a very serious matter. Through short-sightedness the fatal error has been made of disregarding the permanent and actual for the transient and theoretical. If, therefore, we approve such a course of action, we in reality cut the solid ground from beneath our feet. Our view of this matter is fully confirmed by the dissatisfaction of many of our botanists, and their freely expressed intention to use the Rochester Code only until they find something better. Indeed, even one of the leaders among the faithful of late refuses to follow the Code in regard to the starting point for genera.

Of course the greatest fault to be found with the Code arises from the wanton exercise of *ex post facto* legislation to accomplish the ends of its advocates. Were the question of botanical nomenclature in the main a matter of interest to scientists only, as until very recently ornithological nomenclature has been, this legislation would do no great harm if generally assented to; but to employ it in a science like that of botany, where generic and specific names have become, as it were, subjects of property rights, is unwarranted and short-sighted in the extreme. One result has been, that if we follow the Code we may have a botanical name and a horticultural name for the same plant, both correct, but one to be used at one time, one at another, — a somewhat humiliating state of affairs when it is borne in mind what efforts have been made to make horticulturists use the generally accepted botanical names. The worst of the whole matter is that the horticulturists are dealing sensibly with facts as they find them, while the botanists are striving with theories to annihilate facts. It is hard enough, as anybody of experience knows, to make a horticulturist adopt a change in nomenclature made necessary for scientific reasons; but how impossible it would be to force upon him a change made merely to carry out a theory or a system

of questionable expediency! What can be gained by inten-
sifying this distinction between botanical and horticultural
nomenclature, especially now that the horticulturists have
refused to follow the Rochester Code on the practical
ground that it does not recognize the well established
principles of property rights, custom, usage, and the salu-
tary maxim, *Quieta non movere* ?

Any system of nomenclature, especially one creating
confusion by asserting new and unusual theories, should
come before the public as a result of mature, impartial,
long considered adjudication. While we are perfectly
willing to consider the Rochester Code as the expression
of the personal opinion and preference of its advocates,
we find ourselves unable to admit that it has any other
authority to sustain it. It is true that it was fathered
by the vote of the Botanical Section of the A. A. A. S. at
the Rochester Meeting in 1892, but the published record
of the proceedings shows clearly that the committee
appointed could not have given the subject proper consideration and adjudication. In fact, apparently less
than one day was sufficient for this committee to pass
on a subject of so much practical importance, and then
in a manner that involved the rejection of fundamental
principles confirmed and supported for years by the
authority of the greatest botanists. Further comment is
unnecessary.

It has, moreover, been asked, with some pertinence,
What authority had the Rochester Meeting to bind American botanists by any such code of nomenclature as a
majority of the members present might see fit to adopt ?
It is perfectly clear that its sole authority lay in the
united dictation of the various botanists present. We
confess we find it somewhat amusing, — after all the

protest against one-man authority, no matter how great that man might be, and after all the laudation of the democracy of the botanists, — that the real democracy, in which every botanist has a vote, should now be dictated to by a comparatively few botanists of various degrees of repute. History testifies that power and dictation are fully as sweet to thirty tyrants as to one! The matter practically wears this aspect in our opinion, since we have been unable to find more than passive approval of the Code outside of a comparatively small circle of botanists, and in many cases have found active disapproval or a decided disclaimer of any sympathy with the Code where we hardly expected it. We sincerely hope that botanists in other countries will not be deceived into thinking that this school of nomenclature includes the American botanists, for it includes only a part, even if it is the part that makes most of the noise!

Another evil produced by the adoption of this Code is the great prominence given to the botanical name-monger, a term which we use for convenience to denote those botanists who devote much of their time to changing about names of plants for no scientific reason, but merely to fit them to a code. To the binomial thus manufactured they add their names, and stand apparently on a par with botanists whose names attached as authors stand for true scientific achievement. The addition in parentheses of the name of the original author of the specific name does not help the matter much in such cases, for it does not explain the binomial. There are, moreover, no indications at present that there is likely to be such a consensus of agreement in the names of plants as might enable us to omit the name of the author altogether. Thanks to the provincial-mindedness of the so-called reformers, we are farther from this agreement than we were ten years ago. All this is a natural result of the unjustifiable attempt to apply rules too strictly in many respects to the past, over which no botanist can expect to legislate if he knows anything of conditions outside of his herbarium walls. If the supporters of the Rochester Code think they have a right to upset important results of nomenclature evolution for nearly a century and a half merely to help out their theories, they must be veritable Rip Van Winkles, just awakened from a comfortable nap of years.

We sincerely regret that so many of our younger botanists have been led astray by this *ignis fatuus* of theory, and so blinded to the clear fixed lights of sound judgment and of practice. No code of botanical nomenclature can hope to accomplish good results that does not meet the needs of the time; this the Rochester Code does not do. We cannot afford to begin over again, or submit to temporary confusion for the sake of any theory, or for the sake of a future peace that may never come.

In our consideration of this matter we have pointed out a few reasons why we could not follow the Rochester Code. It would be easy to be more specific, and give others, did we feel that it were incumbent on us to do so. We see no reason, however, why objections should be set out by any one dissatisfied with the Code, when the supporters have thus far been unable to prove that it has any right to exist beyond their own will. Let them attempt to prove their case, and their argument will be impartially heard by all interested in this matter of botanical nomenclature. At present they are in default.

In conclusion, we wish to add a few words in explanation of the arrangement adopted in this catalogue. The arrangement of Engler and Prantl's "Natürlichen Pflanzenfamilien," adopted as a standard at the Madison Meeting of the Botanical Section of the A. A. A. S., seems to us not sufficiently well known, accessible, and understood in this country to make it advisable to adopt it at present in a local Flora, or in a mere list of plants. It has been adopted in Algæ alone in this Catalogue. As our decision in regard to the standard of nomenclature obliged us to follow Gray's Manual, it seemed well to us to follow its arrangement also, so far as its scope allowed, and beyond that to follow the manuals and other works already mentioned in the Preface.

AN OUTLINE

OF THE

GEOLOGY OF MOUNT DESERT.

By WILLIAM MORRIS DAVIS.

THE mountain range of MOUNT DESERT includes the highest of a number of mountainous hills that rise over the rolling lowland of southern Maine. The lowland has been slightly inclined to the south, so that a part of its original area is depressed under the sea, to make the platform of the Gulf of Maine; while its northern extension slowly ascends inland until it deserves the name of a plateau in the northern part of the State. The tilted lowland is roughened by the excavation of numerous valleys; and since these were formed the coastal region has been slightly lowered, carrying the shore line farther inland than before, changing many a valley into a long arm of the sea, and isolating many a hill top as an outlying island. Associated with this later change of level, and during a time of colder climate, there was an invasion of the region from the north by a sheet of ice, such as that which still maintains possession of Greenland. The slow but rough-shod march of this cold conqueror stripped the loose soil from the land, wore down the sharper ledges of the hills, deepened many of the valleys, and dragged along the rubbish thus gained farther and farther southward. When the invader was driven away by the return of a milder climate, the rubbish or "drift" was irregularly disposed over the uneven lowlands, thereby greatly embarrassing the flow of the streams that again took possession of the country, frequently turning them aside from their former courses, and often holding them back to form lakes.

The rolling lowland over which the mountainous hills rise is not, like the coastal lowlands of our southern States, a former sea-bottom recently emerged, and still for the most part as smooth as it was when under water. The lowland is low, not because the country was never built up to a greater height, but in spite of having been long ago strongly uplifted in disorderly form. The lowland is low because the whole region has been worn down from its high estate by long continued denudation. It has slowly wasted away under the ceaseless attack of the atmosphere. Its relief is now generally of moderate measure because, before the lowland was tilted into its present southward inclination, it had been worn down nearly to the level of the sea of that time, and only the more resistant rock structures then still withstood denudation successfully enough to hold up their heads as residual mountains and hills. There is every reason to believe that even the residual mountains were once more lofty than they are now; that the whole region was once deformed and upheaved into a rugged highland; but these ancient features have been subdued and almost lost in the denudation of advancing old age. The existing mountains must therefore be regarded not so much as points of excessive upheaval, but as points where the wasting of the land has been retarded. The mountain range of Mount Desert is one of the most stubborn survivors of the ancient highland. The beauty of the Island as seen from the sea, unparalleled along our whole Atlantic coast, is owing to its persistent retention of a good share of the height which this whole region once had, but which its surroundings have lost.

Although the granitic rocks of the Mount Desert range, and of other mountains in southern Maine, are now cold and quiet in their old age, they were once hot and energetic, pressing their way upward as a vast molten mass towards, and perhaps to what was then the surface of the ancient land. Their upheaval and outburst may have contributed largely to the altitude of the former surface; but of this we know little. Other intrusions of melted rocks occurred on large or small scales, and on dates earlier and later than that of the mountain granite; but their heat has all long since died out. The great denudation by which the present lowland has been carved in the ancient highlands is later than the latest of the igneous outbursts; and the glaciation by which the finishing touches have been given to the country is a thing of yesterday.

When a brief summary of geological history is thus presented, the reader, if he is not versed in the interpretation of evidence presented in the language of the rocks, is likely to regard the whole subject as something of a mystery. He may even imagine that the facts and arguments to which the geologist appeals are obscure and abstruse. This is not the case. Common eyes and common sense may perceive all the essential points in the evidence leading to the conclusions just stated. If the reader will walk patiently over the island, look closely, and think clearly, the whole argument may be apprehended; and when his attention is taken less by the conclusions to which the evidence leads than by the evidence which leads to the conclusions, the mystery vanishes; the essential simplicity of logical scientific investigation takes its place, and the face of nature gains a frank and sincere expression that it has to him never worn before. Let us use the foregoing paragraphs in the nature of a table of contents, and now turn more particularly to see

the facts by which the geological history of Mount Desert
may be interpreted.

Seldom are geological facts more plainly presented.
Seldom have pleasanter days been spent than those
recalled while writing out this sketch. We have coasted
in good company and under good pilotage along the rocky
shore, landing for our geological discoveries even as old
Champlain may have landed for his geography, and return-
ing to our vessel at night. We have clambered up pathless
glens to rugged summits; and if we carried rations for
only half a day, we felt nevertheless the spirit of explorers
in unknown lands, and our adventures were recounted
around camp-fires in the evening. Our vacations are
shorter now than then, and while recalling them in this
writing we must leave to others the pleasures on sea and
shore once our own.

THE GRANITE BELT.

The granitic mass of Green Mountain, and of its dominant fellows east and west, and of a belt of adjacent
lowland across the Island about parallel to the range,
serves as a natural beginning in our study, and from the
date of the origin of the granite we may go backwards
and forwards in time until the whole sequence of events
discoverable within our borders is determined. The rocks
of the mountain belt, wherever examined on summits or
flanks, have a remarkably uniform crystalline texture,
consisting of an intimate mixture of quartz, feldspar, and
hornblende, to which the name of hornblendic granite is
given. The constituent minerals may be easily recognized
by the unaided eye: the quartz being translucent and
glassy, with uneven surface; the feldspar, gray or pink,
with even cleavage surfaces; the hornblende, black, and
in smaller particles than the other minerals. The massive
structure of this rock, in so strong a contrast to the bedded

arrangement of stratified or water-deposited rocks, indicates that its materials were not brought here in fine particles and in sucession, and laid down in beds one after another; but that the whole mass took its place essentially at once, and that its structure was gained by a single process, in operation practically at one time in all its parts. Slow crystallization by cooling from fusion is the most plausible explanation of such a result; and this is borne out by an examination of the structure of modern lavas, which solidify after flowing in molten streams from visible vents; and by analogy with the crystallization of mineral substances artificially melted and allowed to cool. The granite is therefore regarded as an igneous rock; a rock which has been at one time molten from heat.

The granite occupies a belt across the Island, enclosed on the north and south by rocks of other kinds. Isolated areas of granite are found eastward from Bass Harbor Head, and at some other points. Descend the mountain slopes to the lower ground, and although much of the surface is covered with drift, the observer will sooner or later meet with rocks of quite different appearance. At first, these are seen as isolated angular fragments of various kinds and sizes included in the granite; the fragments then become more frequent, as in the wonderful display at Hunters Beach Head; further on, the granite is found penetrating long, relatively narrow crevices in the other rocks, as on Sutton Island; and at last the granite ceases entirely, and the surface is occupied, whenever its rocky floor can be seen, only by rocks like those first seen as fragments enclosed in the granite. Near the margin of its area, the granite is finer textured than further within its mass. This indicates that, when it cooled from fusion, the margin cooled faster than the interior; for it is the habit of rocks when crystallizing from a melted state to develop only smaller crystals and finer texture near their boundaries, where they are chilled and solidified quickly; while larger crystals and coarser texture are produced within the mass, where the cooling is more gradual. Even a broken bar of pig iron illustrates a variation from fine to coarse texture in passing from its surface to its interior.

The fine texture of the margin of the granite, the inclusion of numerous angular fragments of the country rocks along the borders of the granite belt, and the penetration of the country rocks by narrowing granitic arms, or dikes, demonstrate that the granite is a later comer than the other rocks, and that it moved from some former position to its present position while molten, breaking its way into the solid, rocky crust in its escape from some excessive pressures that forced it to move; until at last, when the impelling pressures were satisfied, it came to a halt, and slowly froze into a rigid mass, holding close in its grasp thousands of fragments gathered from the enclosing walls. The only imaginable source of supply for such a mass of molten rock is in the earth's interior; and although no one can well account for the forces by which the granite was squeezed outward from its former position, no one can justly doubt the reality of its out-thrust. The granitic belt is indeed nothing more than a great irregular ragged dike, pushed upward through the ancient rocky crust of the earth. Many such intrusive masses are known elsewhere in New England, and in other parts of the world. The Blue Hills near Boston are largely composed of intrusive rocks, amid surroundings much like those of the Mount Desert range; the rocky dome of Cape Ann is of similar nature.

The granitic outburst is the greatest event in the history of Mount Desert. It is of colossal magnitude. The energy of its intrusion cannot be conceived. Not that the intrusion was suddenly accomplished, but that it was effected against enormous resistances, and that it involved the movement of gigantic masses. The granitic belt is at least twelve miles long and seven wide. No one can give any measure of the former greater height to which it ascended ; and certainly he would be a daring geologist who would set a limit to the unsounded depths from which it rose. The uprising may have required many historic ages ; it may have been relatively rapid ; but that it was progressive, and not instantaneous, is easily seen by a closer examination of the minor events recorded along its margins.

Much of the lowland is covered by glacial drift and by postglacial marine clays ; but along the seashore the rocks are swept clean, and their surface is continuously visible. The bare ledges and cliffs of Hunters Beach Head, as well as of many other similar points on the southern coast, afford wonderfully clear illustrations of the processes of the granitic intrusion. Here we may follow the granite wedging its way into narrowing cracks among the older rocks. Great fragments of older rocks of various kinds are caught off in the granite and mixed together in confusion. Sometimes a block is found to be divided by granite-filled fissures, and yet its several parts may still lie so close to one another that they can be matched with certainty ; thus proving that after the block was broken from the wall of the vast fissure it was further fractured, and the minor cracks thus opened were filled by the mobile granite. This may be seen on the eastern side of the Narrows of Somes Sound, and along the shore ledges to Smallidge Point south of Wasgatt Cove. The granite rock, now so rigid, then so liquid, or at least then yielding so perfectly under the pressures that were exerted on it, entered into the narrowest little crevices, following them down to hair-like fineness. Nowhere in the world may the traveller find better illustrations of the manifold processes of deep-seated intrusions than are here exhibited on the wave-swept ledges

4

of the southern coast, eastward from Somes Sound. They
are fully equal to the wonderful display of successive intrusions of igneous rocks along the Massachusetts coast
in Swampscott, Marblehead, and Beverly.

When the observer first examines the varied features of
this rocky coast, his attention will be limited to the surface
of the rocks as now exposed. But the structural problem
of the Island is not simply a problem of surfaces in two
dimensions. It is a problem of solids in three dimensions ;
and the third dimension of height or depth must be inferred
from what can be seen on the length and breadth of the
surface. It will be plain, when we come to consider the
denudation which the island has suffered, that the present
surface has no particular relation to the whole mass of
ancient country rocks and intrusive granite with which we
are now concerned. The present surface merely marks
the stage of denudation reached at this hour of geological
time ; the surface at earlier hours intersected the mass at
a greater altitude ; in later hours the intersection will be
carried lower down into the mass. The present surface
may therefore be taken, not as belonging only to the present
time, but as a fair sample of what would be exhibited on
any nearly horizontal section across the mass, not very far
above or below what is now seen.

We must therefore conceive of the great granite dike
not only as limited by horizontal marginal lines, but as
enclosed by ragged walls ; and this structure must be men-
tally restored upward into what is now the open air, as well
as deeply downward into the solid earth. The original
walls undoubtedly terminated in both directions ; but no
one shall say how far they extended at the time when
the granite had just made its way upwards from the deep
interior of the earth and frozen stiff in its new position.
The greater part of the intrusion is pure unmixed granite ;
but far up and down the walls there must have been a con-

fusion of included fragments broken from the country rock, and a great branch-work of lateral granitic dikes penetrating the sides of the vast fissure. We shall later return to consider the denudation of the ancient mass into its present form; but before that several facts of even more ancient date than the granite intrusion must be examined.

THE PRE-GRANITIC ROCKS.

Even the casual observer can hardly fail to detect a marked variety in the nature of the rock fragments included in the granite along the southern and western coast. Every one of these rocks is older than the granite. Many of them are distinctly unlike in composition and texture, and probably also in age. Some are therefore older than the granite by longer ages than the others. Their sequence must be deciphered as far as possible.*

The most manifest varieties of these older rocks may be briefly described. On the western and northwestern coast, and on some of the adjacent islands, as Bartlett and Hardwood Islands, there is an area of wrinkled greenish schists, in somewhat disorderly attitude, associated with quartzitic layers. Their southernmost occurrence is at Dix Point, and northernmost at Thomas Island. The schists trend northwest, north, and northeast, dipping to the eastward, or towards the granite, at various angles. Their thickness is estimated as two thousand feet at least. These schists are cut by the granite at several points, and hence belong in the pre-granitic series; but as they are not found in contact with the other marginal rocks, it is only on account of their gnarled and ancient appearance that they are placed at the foundation of the history of the Island. They seem

* A number of statements in this section, and in the section on the glacial invasion, are taken from an essay by Prof. N. S. Shaler, on the Geology of the Island of Mount Desert, in the Eighth Annual Report of the Director of the U. S. Geological Survey, Washington, 1889.

to have been originally stratified sedimentary deposits ; but
their texture has become' crystalline by long continued
change, and hence they are to be regarded as extremely
ancient. It is probable that a search on the neighboring
mainland will enable us to define more precisely the relative age of the schists by means of their contact with the
other old rocks ; and when this is done, we may expect to
find proof, not only that the schists are the oldest of the
series, but that after the deposition of their sediments they
were buried deeply enough for metamorphism into their
present crystalline structure, and then greatly denuded
before any other rocks were formed in this region. They
will then be seen to be older than the other rocks by a
great interval of time. The rocks of Schooner Head and
of a limited stretch of the eastern coast may perhaps be
classed with the greenish schists of Bartlett Island; but
their age is not well determined.

On the southern and northeastern sides of the Island,
and on some of the adjacent smaller islands, as Bar Island,
the Porcupines, and Sutton Island, there are many exposures of a series of bedded rocks, partly slates, partly
sandstones and flagstones. These are manifestly sedimentary deposits in ancient seas. They are of firm
texture as a rule, although some of the layers are weak
enough to undermine the overlying beds and form distinct ledges, as near Bar Harbor landing. Sometimes
there are fine pebbly layers, with grains of white quartz,
as on the shore near Northeast Harbor. The greater
part of this series is well indurated ; but otherwise it has
suffered little structural change since its deposition. Indeed, if any fossils had originally existed in the beds, they
should be still observable, but none have yet been found.
The strata generally dip away from the granitic belt; and
on the Cranberry Isles their inclination is nearly vertical.
The granite cuts them in various places, and frequently

includes their fragments along the southern coast; hence, like the greenish schists, they belong to the pre-granitic series, but at present it is only by inference that they are regarded as younger than the schists, as already explained. There may well be a considerable diversity of age among these bedded rocks, yet to be discovered.

The northern shore by the Ovens, the southwestern extension of the Island, and several of the smaller islands to the southeast, contain many exposures of old volcanic rocks, known as felsites. They are of crystalline texture, but much finer than the granite of the central belt, and are often arranged, like modern lavas, in sheets or flows parallel with the beds of the adjacent stratified sedimentary rocks. They frequently possess a porphyritic structure; that is, small crystals of feldspar are disseminated through the mass. Again, they have a banded structure, due to flowing while molten; and they are often broken or brecciated, as if eruptive movement had continued after a part of the mass had become solid. This structure is exhibited on the eastern shore of Bass Harbor. Associated with the denser masses are large fragmental and ash-like deposits, as if formed by explosive eruptions from some neighboring vent not now identifiable. Occasionally, dikes of felsite are found cutting through the rocks of the bedded series. Like the bedded rocks, they are cut by the granites, as may be well seen east of Bass Harbor. Considering all these features, it may be concluded that the felsites mark a time of volcanic activity contemporaneous with a part of the period during which the sedimentary series was formed. The sandy or muddy sea bottom of that era must have received, from time to time, flows of lava and showers of ashes; volcanic cones may have been built somewhere in the neighborhood, although not a trace of them now remains. While in process of accumulation, the bedded rocks and the lava flows must have lain almost horizontal;

but they are now steeply tilted to the south and deeply worn away, so that the present surface in the Cranberry Isles reveals what would have been originally almost a vertical cross-section of the mass. Minute study may discover interesting details of this chapter of the Island's history ; but the record is fragmentary by reason of the denudation that has swept much of the structure away, and the submergence which has sunk a good part of the remainder beneath the sea; and the remnant standing above the present sea level is blurred over by the sheet of glacial drift. Yet it is by putting together such imperfect records as this that much of the geological history of the world has been made out. A close study of the ancient volcanic area in the southern part of Mount Desert would doubtless well repay any one who can undertake it.

Along the southern coast, east of Somes Sound, and at various points on the western coast near Bartlett Island Narrows, there is a dark-colored crystalline igneous rock, known as diorite. It consists chiefly of hornblende and a triclinic feldspar : the fine parallel lines on the cleavage faces of the feldspar resulting from the twinning of crystals can be easily seen with a hand lens, thus distinguishing the triclinic feldspar of the diorite from the orthoclase feldspar of the granite. The diorite is shown to be an igneous rock by its structure, and by its intrusive relations with other rocks, here and elsewhere. It cuts the lower members of the stratified rocks near Northeast Harbor, and it is frequently cut by or included in the granite ; hence its age is intermediate between the ages of these two ; but its relations to the volcanic felsites are not yet surely determined. The felsites were contemporary superficial extrusions upon certain members of the bedded rocks, while the diorite presents only the features of a deep intrusion, as if thrust in among the bedded rocks after they had accumulated in much greater mass than they possessed when the felsites overflowed. It is there-fore probable that the diorite is younger than the felsite. At certain points in Northwest Cove, on the western coast, the diorite is of two kinds; the finer textured masses being cut by those of coarser texture, and thus indicating two periods of intrusion of this rock.

This general survey of the older rocks may now be summarized. The metamorphic schists seem to be the most ancient, and it is probable that a long unrecorded time elapsed after their deposition before the next series was formed. Then comes a variety of unaltered sedimen-tary rocks, to whose accumulation a long time must have been devoted, and whose history was diversified by much volcanic action. The geological date of these rocks can-not be affirmed ; but, judging by analogy with similar rocks along the New England coast, it was probably Cambrian, a very ancient time division of geological history. How long the conditions of deposition prevailed, and how great a thickness of deposits was formed before their subsequent destruction began, no one can now learn ; but we shall see reason to believe that the existing amount of bedded rocks is probably only a small share of what once existed in this region. At some part of the time, when the accumulation of the bedded series had reached a considerable volume above the present surface, the intrusions of diorite took place within the mass ; and again, at a still later time, came the great granitic intrusion ; thus leading us to the undated epoch with which this account of the rocks began.

THE POST-GRANITIC ROCKS.

The only indurated rocks now recognized as of later date than the granite are the trap dikes, by which all the other rocks of the Island are traversed at one place or

another. These dikes are commonly from two to ten feet
wide, standing nearly vertical, trending somewhat east of
north with rather direct courses. They are found on lowland and highland, from water's edge to mountain top.
They nowhere exhibit the smallest indication of overflow;
even over the summit of Green Mountain they are as dense
and as well contained within their walls as at sea level.
Hence, when they were intruded, the rocky mass must
have risen above the mountain tops of to-day. It is possible that their lavas may have reached the surface of
their time, and may have there overflowed, much in the
same way as the felsites overflowed on a lower surface
long before ; but we have no evidence of such surface
action. The dikes, as now revealed, are deep structures,
and with the diorites and granites proclaim the greater
mass that the rocks once possessed, and the wasting that
the region has since then suffered.

THE GREAT DENUDATION.

If the greenish schists are older than the bedded rocks,
as may be supposed, it is eminently possible that the
unrecorded time between the deposition of the two series
witnessed a denudation as great as that which we have
now to consider ; but our attention cannot be well directed
to that long lost chapter of the Island's history. The
chapter is at present only a matter of fair inference, and
it never can be fully reconstructed. It is like those many
lapsed periods of ancient human history, unmarked by
records of great battles or by the dethronement of kings,
over which our imagination passes so lightly, and with so
little appreciation of all that they contained. But with
the later denudation, by which the present form of the
Island has been fashioned, we have much to do. It calls
for study as attentive as that by which the making of the
rocks is discovered. A full understanding of the geological

history of a region requires an examination of the records of denudation, as well as of those of accumulation. It is likely that Nature gave quite as much time to wearing down as to building up the Island; and we may well follow her example in the division of our sections.

The reasons for believing that the Island and the adjacent mainland have lost much of the rock mass that once existed above their present surface may be briefly stated. We first notice that, if the tilted beds of the stratified and volcanic series were again extended upwards into the air from their present denuded edges, a great increase would be given to the altitude of the surface. This is not merely a local matter. The same conclusion is reached all along the New England coast, and far inland. The rocks of the whole region are greatly disordered, much as rocks are in lofty mountains, and the edges of the strata as now revealed are by no means the original edges. How far they once extended upwards cannot be stated; but the distance should be estimated in thousands of feet rather than in hundreds.

The evidence thus derived from the attitude of the bedded rocks is confirmed by the features of the intrusive rocks, — the diorites, the granite, and the trap dikes. None of these exhibit any trace of surface extrusion, such as is so plainly manifested in the more ancient felsites. Hence we must suppose that, since the felsites were extruded, a great accumulation of superincumbent materials was loaded upon the region, and that it was upwards into this heavy accumulation that the intrusive rocks were thrust. It is quite probable that during the time of accumulation and intrusion the whole region stood lower than it now does, even so low that its surface then was near or below sea level. As long as this low stand was maintained, further accumulation would be natural enough, and denudation would be postponed. No limits of quantity or time can

now be placed to the era of accumulation ; but it was closed
at last by elevation, and with that change the present chapter in the history of the Island was opened. Judging by
the tilted and twisted attitude of the bedded rocks, both on
the Island and elsewhere in New England, it is probable
that the time of elevation was a time of mountain growth,
when the rocks were deformed as well as uplifted. The
coast line must then have been pushed farther out to sea,
nearer to the margin of our continental shelf. The mountains may have risen as high as the Alps ; they may have
borne glaciers on their upper slopes ; great rivers may have
drained their valleys. The rocks may have suffered mountainous deformation at more than one period, writhing
under successive applications of crushing forces, after the
fashion of mountains of more recent construction, whose
building is better known. During, between, and after the
periods of crushing, the forces of the atmosphere maintained
their ceaseless attack on the exposed surface ; and their
final success in reducing the ancient mountains so nearly
to a lowland reminds one that the persevering tortoise over-
took the spasmodic hare.

The lack of definiteness by which this section is charac-
terized may make appreciation of it more difficult than
of one which, like the section on the intrusion of the
granite, is accompanied by specific illustration at every
step. All the more patiently, therefore, should the reader
pass in review the scenes of existing mountains, having
faith that where mountain roots are now exposed, there
mountain heads once arose. Just as he might recall, while
resting on the prostrate trunk of an old moss-covered
forest tree the early sprouting of its seed, its adolescent
growth above the lowlier bushes, its mature attainment of
forest height, its fall and decay in old age : so he must
picture the young mountains once rising along what is now
the New England coast ; he must see them grow once and

again, rearing their crests to the height of the clouds; he must watch them slowly, slowly wearing away till only their roots remain. Even as we are told that in man death begins with birth, so with the mountains: their wasting begins as they first rise above the sea, and, however lofty they grow, they must in the end be prostrated. Energetic mountains of great altitude, so young that they are still high and growing, are not the only kind of mountains that cross the face of the earth. Many a mountain range, once lofty, has been laid low; and it is as a part of such a range that we must regard both the highlands and lowlands of Mount Desert.

The narrow limits of the Island suffice to give us an understanding of its granitic belt; but in the present section the Island must be regarded merely as a part of New England. It is only from a general survey of a considerable area that a just view of its parts can be gained. Let the reader, therefore, now recollect what he has seen of New England elsewhere, and follow a rapid sketch of its history as a wasting land.

Although New England is a rugged country, an extended view from its hill tops brings to sight a comparatively even sky line, at whose elevation extensive uplands often stretch many miles without great inequality of height. The mountains that rise above this sky line and the valleys that sink below it may be for the moment left out of consideration. The upland is the most general feature of inner New England, and must be distinctly recognized. This upland surface, gradually descending towards the seacoast, merges into the lowland that was mentioned in the first sentence of this essay. The whole is manifestly a surface of denudation, for its stratified rocks are nearly always exposed on edge, and their former extension has been greatly curtailed; while its crystalline rocks in almost every case possess a coarseness of texture and a structural relation to

their surroundings that indicate intrusion at great depths
beneath the surface of their time. Now it is a law, well
demonstrated in the science of land sculpture, that an even
upland of large extent, like that of the New England plateau, in which there is at present no sympathy between
rock structure and surface form, can be produced only in
the later stages of a long cycle of denudation, when the
region has wasted from whatever height it once possessed
nearly down to sea level, or base level as it is conveniently
called. The New England upland was therefore once, not
only at its margin as now, but across its whole extent, a
lowland of denudation standing near sea level; and its
present elevation must have been given to it at some subsequent time by an unequal tilting which depressed part
of its former extent beneath the sea, and which raised the
inner portion of its area one or two thousand feet.

The reader must guard against making too even a
picture of this ancient plain of denudation. It was by
no means a dead-level plain, but a rolling surface of moderate relief, — an almost plain surface, for which I have
coined the term *peneplain*. During the long cycle of
denudation, the region was not entirely worn down to sea
level, but it was greatly reduced from the height that it
once possessed, and only low hills remained to represent
most of its ancient mountains. At certain points, not
even the peneplain stage was reached. The view across
the New England upland nearly always includes, in one
direction or another, an eminence rising above the general
sky line; and of such, Monadnock in southwestern New
Hampshire is one of the most beautiful examples. These
eminences, once overlooking the lowland, but now overlooking the plateau, are residual mountains, which by reason
of their excessive hardness escaped the nearly complete
denudation that the rest of the peneplain suffered. The
White Mountains seem to be simply a cluster of Monad-

nocks. The range of Mount Desert is a series of Monadnocks, close to the line where the sea now lies on the land.

It is only in a later cycle of denudation, since the lowland peneplain was uplifted into a rolling plateau inclining towards the sea, that the valleys of New England were carved out. Where the plateau rose high, the valleys have been cut deep; where it rose but little, there are only shallow trenches in the upland. Where the rocks of the uplifted peneplain are relatively hard, the valleys are as yet, in the present cycle of denudation, opened only to a moderate width; where the rocks are weak, the valleys are opened so wide that new local lowlands, local peneplains of the second order, or of a second generation, have been developed. Thus the rocky floor of New England is diversified, and of this rocky floor Mount Desert is a little part.

Just as we must avoid too artificial a conception of the plain to which New England was reduced by the great denudation, so we must guard against too rigid an assumption of a perfect standstill of the land during either the greater or the lesser cycle of its degradation, and too violent an assumption of its immediate ascent from a lower to a higher altitude. It is most probable that many minor oscillations of level occurred during the cycle while the peneplain was in development, and that the elevation and tilting of the peneplain into the inclined upland was so gently accomplished, that, had we then been here to watch its rise, we might have watched in vain during our too brief centuries. It is also probable that, since the uplift and the beginning of the new cycle in which the valleys have been etched out, minor oscillations have again upheaved and depressed the land. It is only the greater changes of level that can be detected in the more remote history of the sculpture of the land; it is only as we come close to the present time that the minute records of slight and brief oscillations can be detected. The moderate depression by which our lower valleys have been, as we shall see, drowned into bays, and the lesser elevation by which our coastal slope has risen with a half-smooth sea cover on its back are important to us, not by reason of their magnitude or their duration, but simply by reason of their recency. They must not be regarded as exceptional, but only as giving indication of the uneasiness that most likely always has and always will characterize the land.

There is nothing on Mount Desert, or on the coast of Maine, that suffices to define the geological date of the elevation by which the two cycles of denudation just described were separated; yet when the field of inquiry is extended so as to include all parts of the uplifted peneplain, which is found to spread far to the southwest, even to Alabama, its denudation may be correlated with the deposition of various fossiliferous sediments, and thus the completion of the peneplain may be placed in its proper position in geological chronology. Strata of late Cretaceous age in New Jersey are found overlapping the seaward margin of the peneplain; hence it is believed to have been fairly well completed in late Cretaceous time; and the period of its elevation and consequent etching is regarded as post-Cretaceous, or somewhere in the Tertiary period. This is manifestly rather indefinite; future investigations will probably define it more sharply; but it is a significant step in the right direction. Before this small step was made, the date of the denudation of New England was entirely unfixed, and very diverse views were held on this subject. The making of the peneplain was by some thought to be as old as the red sandstones of the Triassic formation in the Connecticut valley; and the valleys were considered by others to be as young as the time of the ice invasion, to whose erosive powers they were ascribed. The truth as it now appears lies between these extremes.

The geological dates of the intrusions and deformations that the Island has suffered are even more indefinite than the times of its denudation. Accepting the provisional date of Cambrian for the lower members of the bedded series, and late Cretaceous for the end of the denudation of the peneplain, the deformations and intrusions must be placed somewhere within the long intermediate interval. This is like saying that a certain battle occurred somewhere between the time of the founding of Rome and the invention of printing; but if its date had been previously still less determinate than this, we should be glad enough to have even so wide a limitation of its occurrence. It is probable that comparisons of the structure of Mount Desert with similar structures in other parts of New England will in future suffice to set narrower limits to the dates of several of the events whose time of occurrence is now so loosely circumscribed.

THE GLACIAL INVASION.

It was over a country thus made and unmade, over an uplifted peneplain surmounted by isolated and clustered Monadnocks, and dissected by newly etched valleys and valley lowlands, that the ice sheet crept down from the north. I shall not enter on the cause of its coming, or on its source; suffice it to call attention to the manifest marks of its presence. Wherever the rocky floor of the Island, or of almost any part of New England, is freshly uncovered, it is found to be more or less smoothed or rounded and distinctly striated or grooved, as if it had been severely rubbed down by some gritty burnisher. Such surfaces may be seen at innumerable points along the rocky shore of Mount Desert, where the drift has lately been stripped off and where the waves have not yet made successful attack. On the mountain tops the striæ are weathered off, but the rounded form of many ledges is a significant product of glacial action. The striæ trend to the east of south, sometimes deflected to one side or the other by the uneven form of the hills and mountains, but generally persisting rather regularly in their course. These markings are so perfectly matched by those seen on the rocks under the creeping ice streams of Switzerland, or alongside of the decaying ice sheet of Alaska, that it is unreasonable to doubt that they were produced in the same way. Mount Desert, Maine, New England, all our northern States and Canada beyond them, are thus engraved; over all of them once lay an ice sheet at least twice as large as that which now swathes Greenland; and this so little time ago that the ice scorings are still fresh, where protected only by a thin layer of drift. The ice moved outwards towards its margin; and in Maine this was to the south-southeast. We must imagine it advancing beyond the present coast line, and terminating in an ice wall in the sea, yielding innumerable icebergs to float for a time southward in the cold 'long-shore current.

There is a curious transportation of boulders that must be associated with the glacial scoring of the rocks, for it runs in the same direction, and it is of too great an amount to be of postglacial date. It is true that on the young sea cliffs, and at the foot of many of the steeper ledges on the mountain sides, the wasting of the rock progresses rapidly, and blocks have been loosened and moved over short distances, or have fallen down the talus slopes, since the ice sheet retreated; but so local a distribution of rock fragments will by no means account for the long carriage of the innumerable boulders that are scattered far and wide over the country. Boulders of the easily recognized greenish schists of the western coast of the Island are found removed southward from their parent ledges, in areas underlain by rocks of other kinds. Boulders from the granite of the central belt are moved southward in plenty over the surface of the bedded and volcanic rocks. Not only so; blocks of a coarse gray granite, easily known by its large crystals of whitish feldspar, but not occurring in the rocky structure of the island, are found here and there over its surface. They come from the mainland, where this kind of granite is well known. One of these boulders was to be seen, some years ago, close to the summit of Green Mountain. Further than this, there are fragments, generally of less than a foot in diameter, of a fossiliferous shaly flagstone, sparingly distributed over the western half of the Island; these are easily identified as belonging to a belt of Devonian strata some miles northward from the mainland coast. No such rocks occur in place on the Island. All of this peculiar transportation of erratic boulders is ascribed to the ice sheet, aided in some cases by its subglacial streams. The boulders are simply the larger fragments that the ice sheet dragged along beneath it, or carried in its lower portion.

The unconsolidated drift by which the lower rocky floor of the island is generally covered frequently possesses a structure that gives still further indication of land-ice action. Its lower part, lying close packed on the striated bedrock, is a compact unstratified mass of stones, sand, and clay; the stones are of both local and distant origin, being more worn and striated if from a distance, while those that have been brought but a little way show fewer signs of severe usage. Deposits of this kind are called by the Scotch name, *till*. They are very generally spread over the New England plateau, where they diminish the ruggedness of the rocky surface. Till also occurs in the valleys; but here it is often covered over by water-washed sands or clays of somewhat later date. On the " Monadnocks," the till is scanty: above five or six hundred feet, the Mount

Desert range exposes a large surface of bare rock. In many parts of New England, the till is accumulated in large rounded hills, of oval outline and smoothly rounded profile, called by the Irish name, drumlins. These are common in the neighborhood of Boston, and further inland about Brookfield and Pomfret; but with half an exception they are absent on Mount Desert. This half exception is the long smooth northern slope of Beech Hill, southwest of Somesville; apparently a deposit of till simulating a half-drumlin form, extending only northward from the rocky knob of the hill summit; while to the south, where a completed drumlin would descend symmetrically, there is a rocky slope.

Under the ice and in its lower part many blocks were moved from their native ledges; the preglacial soil was scraped off, and the rock beneath was rubbed down. Valleys were deepened and hills were degraded; but by comparing regions inside and outside of the glaciated area, it is plain that as a rule no great erosion of hard rocks must be attributed to glacial action. The excavation of our valleys in the uplifted peneplain was a large piece of work compared to the scraping of the surface by the ice sheet; and the time required for the valley making was much longer than the duration of the ice invasion. Yet on Mount Desert there are certain considerable topographic features whose origin has no other explanation than excavation by the rough-shod ice. These are the deep transverse valleys by which the mountain range is so curiously divided. In its moderate length of twelve miles, it is notched almost down to or beneath sea level no less than nine times. Instead of a mountain ridge as continuous as the granite of the central belt, we have a beautifully diversified succession of rounded domes, separated by deep gorges; and in nearly every gorge there is a lake or an arm of the sea, almost directly in the axis of the range. There is no understanding of this exceptional form, unless it can be explained as a glacial modification of a mountain range previously serrated by transverse notches of moderate depth, down from which lateral ravines descended to the lowlands, north and south. It has therefore been supposed that the more rapid flow of the ice through these preglacial passes gouged them out as deep as the open lowland on either side, or even deeper.

There are few well marked examples on Mount Desert of the curious deposits of gravel and sand, elsewhere common enough in New England, that were formed during the closing stages of the ice period. Many of the mainland valleys are half clogged with heaps, ridges, or plains of these loose materials, lying upon scored rock or upon a varying sheet of till; their origin being ascribed to streams that ran from the waning ice sheet, discharging their load of sand and gravel in the open spaces along the ice border. The absence of such gravels on Mount Desert would imply that during the disappearance of the ice sheet the streams from the mainland avoided the island, and followed by preference the lower districts east or west. The surface deposits of drift are, however, frequently of a gravelly nature, especially at altitudes above that of the clays later described, and below elevations of four to six hundred feet; and these may be ascribed to the wash of streams and currents as the ice melted away. Above the head of Bass Harbor, they attain an uneven form, characteristic of the gravel mounds or kames of the mainland; but this is exceptional on the Island.

POSTGLACIAL HISTORY.

What with the deepening of the transverse valleys and the irregular deposition of the till over the rocky floor, we find the drainage of the Island peculiarly embarrassed since the disappearance of the ice. In preglacial time,

when the drainage lines had been for a long time under
control of sub-aerial streams, it is most probable that there
were no lakes on the Island in which the streams were
detained on their way to the sea ; now there are twelve
lakes, several of them of good size and most picturesquely
placed amid the mountains. If the land stood a little
higher, Somes Sound would be transformed into a lake,
for its waters are deeper in the line of the mountain range
than farther south at its Narrows. It is interesting to
notice that the streams which enter Denning Pond (Echo
Lake) and Great Pond come from the south, the head-
waters of the latter lying distinctly beyond the axis of the
range ; and that the northward outlets of these lakes lead
their waters into the head of Somes Sound, through which
they pass back again southward across the line of the
range to the sea. Such an arrangement of streams would
be unnatural or impossible in a region whose drainage
had been developed under the ordinary processes of at-
mospheric wasting ; and must be, with the occurrence of
the lakes deep in the mountain axis, referred to glacial
action. It is noteworthy, however, that while the rock
scorings, the transported boulders, and the till are all
taken as demonstrating the existence of an ice sheet in
the recent past, the occurrence of the lakes and of the
reversed or northward drainage is not generally regarded
as belonging in the same demonstrative category. The
latter facts may be plausibly explained by the action of
the ice, the existence of which is to be otherwise demonstrated, rather than regarded as independent indications
of ice action, even in the absence of other evidence.

The altitude of the land at the time when the ice invasion began is not known, except that for a considerable
preglacial time it must have been somewhat higher than
now, to allow the excavation as land valleys of the many
arms of the sea that now break up the coast line of Maine.

But at the closing stages of the ice invasion and for a time afterwards, the land must have stood lower than now; for beds of stratified clay bearing marine fossils are found at various points on the lowland of the Island and the mainland, up to about two hundred feet above the present sea level. Judging by the relation of these clays to the washed gravels and sands on the higher slopes, it is probable that the submergence about the close of glacial time amounted here to at least three hundred feet. The clays are relatively scanty on the southern side of the Island ; they are exposed above the shores of Wasgatt Cove, near the Narrows of Somes Sound, and in Seal Harbor. On the northern lowland, they have a wide extension, and conceal the rocky floor over much of the district. On the mainland, the clays and sands are so plentiful over the coastal lowland that they greatly diminish the ruggedness of the rolling foundation on which they lie; and they efficiently aid the deposits of till in displacing the rivers from their preglacial courses.

During the depression indicated by these stratified deposits, Mount Desert was not a single island, but a row of imperfectly connected mountains. Somes Sound was then a thoroughfare, and had several fellows on either side. Nearly all the lower stretches of the Island were submerged. Not only so; at that time the scanty remnants of rocks other than the granite, now visible in patches along the shore, must have been entirely concealed beneath the sea. The geological structure then visible would have been extremely simple.

The depression of the land about the close of the glacial period cannot have been maintained long at any one level, for nowhere on the slopes of the island are there shore lines of as great distinctness as those which mark the present margin of the sea. A depression of much more than three hundred feet has been inferred by Professor

Shaler, but if it amounted to as great a measure as he concludes it must have been of brief duration, as its records are indistinct.

At the present time the land has but partially recovered from the late glacial submergence. Many of the preglacial valleys and valley lowlands are submerged as sounds and bays, and the coast line is probably at least twenty or thirty miles farther inland than it was in preglacial time. It is for this reason that Mount Desert is isolated from the mainland, and that the many other islands fringe the coast. All of these were once hills on a coastal lowland, and when thus exposed there would have been better opportunity than now of discovering the true history of the pre-granitic rocks.

While in its present attitude, the sea has begun to make its mark along the shore. As is the habit on steep coasts of hard rock, the waves excavate caves wherever the rate of cutting on the water line at the base of the slope is faster than the wasting of the slope above; but it is seldom that this relation is found unless aided by joints or other lines of structural weakness in the rocks near sea level. Generally the wasting of the face of the slope on young shore lines about keeps pace with the undercutting of the waves at the base; and thus a rocky bench is formed a little below water level, surmounted by such vertical faces as Great Head and Otter Cliffs. A considerable part of our rocky shore is benched in this manner, but less emphatically. At other parts of the shore, where the land slope is more gentle, yet well exposed to inrolling surf, the loose rocks gathered from the adjacent headlands, and carried in by storm waves from the shelving bottom, are thrown along the water's edge a little beyond high-water mark, making a sea wall, such as occurs in a re-entrant on the shore south of Southwest Harbor. In more protected situations, the embankment formed by the

waves consists only of cobble stones, or of gravel and
sand, as at various points along the western coast. The
embankment may be built across the mouth of a bay, thus
enclosing Long Pond from Bracy Cove, east of Northeast
Harbor; or it may stretch out and tie an island to the
shore, as at Bar Harbor. Bars of this kind are better
developed on sandier and shallower shores than on the
steep and broken coast of Mount Desert.

As the waves rise and fall in broken rhythm on the
shore, as the tide flows and ebbs across the littoral belt,
so the seas of former times have risen and fallen in
uneven measure on the uneasy land; the rocks have
grown and wasted; the ice of the North has crept down
and melted away; — all shifting back and forth in their
cycles of change. Only one scene lies before us of the
many that have floated through the past.

FLORA OF MOUNT DESERT.

CATALOGUE OF PLANTS.

Class I. DICOTYLEDONES ANGIOSPERMEÆ.

Division I. POLYPETALÆ.

RANUNCULACEÆ. Crowfoot Family.

CLEMATIS, L. Virgin's Bower.

C. Virginiana, L.

Thickets; infrequent. Wasgatt Brook; mouth of Hadlock Brook (J. L. Wakefield); — Northeast Harbor, etc. (Rand); — Duck Brook (Rand, F. M. Day); — on Doctors Brook (R. & R.); — Echo Notch (Redfield).

ANEMONE, L. Wind-flower.

A. nemorosa, L.

Rare; apparently occurring only in the southeastern part of the Island. Otter Creek (Grace H. Eliot); — near Schooner Head (Clara L. Walley, Mary Minot).

THALICTRUM, L. Meadow Rue.

T. polygamum, Muhl. *T. Cornuti*, Man., 5th ed., *non* L.*
Common in wet ground, by streams, etc.

RANUNCULUS, L. Crowfoot. Buttercup.

R. Cymbalaria, Pursh. Seaside Crowfoot.
Sandy or muddy shores on the coast; common.

* See Trelease in Bot. Gaz., xi. 92.

R. Flammula, L., var. **reptans** (L.), E. Meyer. Creeping
Spearwort.
Gravelly shores; infrequent. Pool near Schooner Head (Red-
field, G. Hunt); — shores of Great Pond; Ripples Pond (Rand).

R. abortivus, L. Small-flowered Crowfoot.
Grassy fields; rare. Southwest Harbor; Somesville; Bar
Harbor (Rand). Probably introduced in grass seed.

R. recurvatus, Poir.
Rare. Hadlock Valley (Redfield).

R. repens, L. Spotted-leaf Buttercup.
Frequent in moist ground by roadsides, in meadows, etc.
While in some places, as by roadsides, this species is doubtless introduced, it appears indigenous in others. It is not
so abundant on the Island as to lead to any strong presump-
tion of its general introduction, or of its spreading to some
remote places.

R. acris, L. Tall Buttercup.
Common everywhere. Naturalized from Europe.

COPTIS, Salisb. Goldthread.
C. trifolia (L.), Salisb.
Common in damp woods.

AQUILEGIA, L. Columbine.
A. Canadensis, L.
Rare. High Head (Rand); — East Point, Seal Harbor (A.
Cope).

A. vulgaris, L. Garden Columbine.
Occasional by roadsides and in waste places. Escaped from
cultivation. Near Jordan Pond (Rand); — near Ovens (Greenleaf, Lane & Rand) ; — Town Hill (R. & R.).

ACTÆA, L. Baneberry.
A. alba (L.), Miller.
Deep woods; infrequent. Sargent Mt. Gorge, etc. (Rand); —
Hadlock Valley (Redfield); — near Bar Harbor (F. M. Day).

BERBERIDACEÆ. Barberry Family.

BERBERIS, L. Barberry.

B. vulgaris, L. Common Barberry.

Escaped from gardens, or rarely spontaneous. Roadside north of Seal Harbor,— this station now destroyed (Rand, Redfield);— clearing, Canada Valley (Rand);— roadside near Norwood Cove (Rand, Annie S. Downs); — Ox Hill, Seal Harbor (Redfield).

NYMPHÆACEÆ. Water-lily Family.

BRASENIA, Schreb. Water Shield.

B. peltata (Thunb.), Pursh.

Abundant in Witch Hole (Rand, Redfield, F. M. Day);— Somes Pond (R. & R., M. L. Fernald); — Ripples Pond (Rand).

NYMPHÆA, L. Water-lily.

N. odorata, Ait.

Common in ponds and meadow streams. A form with very small flowers, Mountain Pond, Sargent Mt. (Rand).

NUPHAR, Smith. Cow-lily. Spatter Dock. Yellow Pond-lily.

N. advena, Ait. f.

Common in ponds, slow streams, and bog holes.

SARRACENIACEÆ. Pitcher-plant Family.

SARRACENIA, L. Pitcher-plant. Side-saddle Flower.

S. purpurea, L.

Common in peat bogs.

FUMARIACEÆ. Fumitory Family.

CORYDALIS, Vent.

C. glauca (Mœnch), Pursh.

Rocky ground and burnt clearings; frequent, but nowhere in great abundance.

FUMARIA, L. FUMITORY.

F. OFFICINALIS, L.

Waste ground, Great Cranberry Isle (R. & R.).   Adventive from Europe.

CRUCIFERÆ. MUSTARD FAMILY.

CARDAMINE, L. BITTER CRESS.

C. hirsuta, L.

Frequent in brooks and on pond shores. So far as known the Island plants are all glabrous, and in other respects do not correspond to the typical European plant. Whether our common American plant is not specifically distinct seems to be an open question. That it is so distinct, see N. L. Britton, Bull. Torr. Bot. Club, xix. 219. As, however, there appear to be intermediate forms, perhaps it would be wiser to give it only varietal rank. The Mt. Desert forms can perhaps be classified under the three following heads for convenience, although apparently there are no well defined dividing lines between them.

(*a*) *Forma* Pennsylvanica. *C. Pennsylvanica*, Muhl. Glabrous; large and leafy; few, if any, radical leaves; pods linear; pedicels somewhat divergent. Roadside ditch between Town Hill and Northwest Cove; near outlet of Great Pond; Intervale Brook (Rand).

(*β*) Like the last, but with widely divergent pedicels, and thicker, much shorter pods. The most common form. Brook, Clark Valley; Cold Brook; Intervale Brook (Rand); — Doctors Brook; Stanley Brook (Redfield); — Deer Brook (R. & R.).

(*γ*) A form more nearly corresponding to typical *C. hirsuta*. Glabrous; radical leaves rosulate; pedicels erect or somewhat spreading; style short and stout; pods variable in length and thickness. Shores of Northwest Arm, Great Pond (Rand, M. L. Fernald).

C. parviflora, L.

Leaflets mostly linear; radical leaves few or none; pods linear, erect on spreading pedicels. In dry ground, or among moist rocks; rare. Little Duck Island; Flying Mt. (Rand).

NASTURTIUM, R. Br. WATER CRESS.

N. palustre (Leys.), DC. MARSH CRESS.
Rare. Field, Somesville (Rand). Doubtless introduced.

N. ARMORACIA (L.), Fries. HORSERADISH.
Rare. Escaped from cultivation to waste places. Somesville (Rand). Adventive from Europe.

BARBAREA, R. Br. WINTER CRESS.

B. vulgaris, R. Br. YELLOW ROCKET.
Rare. Wayside, Bar Harbor (Rand). Lately introduced.

SISYMBRIUM, L. HEDGE MUSTARD.

S. OFFICINALE (L.), Scop. COMMON HEDGE MUSTARD.
Roadsides and waste places. Naturalized from Europe.

BRASSICA, L.

B. SINAPISTRUM, Boiss. CHARLOCK.
Old fields and waste places ; infrequent. Northeast Harbor; Southwest Harbor; Somesville (Rand). Adventive from Europe.

B. NIGRA (L.), Koch. BLACK MUSTARD.
Old fields and waste places; frequent. Northeast Harbor; High Head; beach, Greening Island; Southwest Harbor (Rand). Adventive from Europe.

B. CAMPESTRIS, L. TURNIP.
Old fields and waste places; frequent. Somesville (R. & R.); — Seal Harbor (Redfield); — Southwest Harbor; Northeast Harbor (Rand). Introduced from Europe.

CAPSELLA, Medic. SHEPHERD'S PURSE.

C. BURSA-PASTORIS (L.), Mœnch.
A common weed, — even at Duck Islands (Redfield). Naturalized from Europe.

LEPIDIUM, L. Peppergrass.

L. Virginicum, L.

Roadsides ; rare. Southwest Harbor (M. L. Fernald); —
Somesville (Rand); — Bar Harbor (W. H. Manning). Recently
introduced from the South.

CAKILE, Gærtn. Sea Rocket.

C. Americana, Nutt.

Common on sea beaches.

RAPHANUS, L. Radish.

R. Raphanistrum, L. Wild Radish. Jointed Charlock.

Old fields and waste places; frequent. Northeast Harbor
(W. H. Dunbar); — Southwest Harbor; Beech Hill; Somesville;
Bar Harbor, etc. (Rand). Adventive from Europe.

CISTACEÆ. Rock-Rose Family.

HUDSONIA, L.

H. ericoides, L. Heath-like Hudsonia.

Frequent on mountain summits. Also on borders of Sea Wall
Swamp, and on Bass Harbor road (Annie S. Downs).

LECHEA, L. Pinweed.

L. minor, L., var.

Very common in dry soil. The form found on the Island cor-
responds to *L. intermedia*, Leggett MS. = *L. Leggettii*, Britt. &
Holl., var. *intermedia* (Legg.), Britt. & Holl., — according to
Dr. N. L. Britton.

VIOLACEÆ. Violet Family.

VIOLA, L. Violet.

V. palmata, L., var. cucullata (Ait.), Gray.

Common, mostly in moist ground. Very variable.

Forma **albiflora.**

Flowers pure white. Occasional. Emery District (Wm. C.
Lane).

Forma **variegata.**

Flowers blue, mottled with white. Occasional. Southwest Harbor; Somesville (Rand).

V. sagittata, Ait. ARROW-LEAVED VIOLET.

Frequent in open dry ground,—pastures and hillsides.

V. blanda, Willd. SWEET WHITE VIOLET.

Common in wet places.

Var. **renifolia,** Gray. KIDNEY-LEAVED VIOLET.

Occasional. Southwest Harbor (Greenleaf, Lane & Rand); — old Beech Hill road, head of Norwood Cove (Rand).

Var. **palustriformis,** Gray.

Not uncommon in mossy ground. Seal Harbor (Redfield); — Southwest Harbor (M. L. Fernald).

V. primulæfolia, L. PRIMROSE-LEAVED VIOLET.

Infrequent. Southwest Harbor (Greenleaf, Lane & Rand, M. L. Fernald); — old road to Beech Hill, head of Norwood Cove; shore of Pond Heath (Rand).

V. lanceolata, L. LANCE-LEAVED VIOLET.

Common in wet places and roadside ditches.

V. canina, L., var. **Muhlenbergii** (Torr.), Gray. DOG VIOLET.

Rare. Bar Harbor (Margaret A. Rand).

Var. **puberula,** S. Watson.

Frequent in dry soil. High Head; pasture near Pond Heath; Northwest Cove (Rand); — Bar Harbor (Mary Minot).

V. TRICOLOR, L. PANSY. HEART'S-EASE. LADIES' DELIGHT.

Escaped from cultivation. Bar Harbor (W. H. Manning). Adventive from Europe.

CARYOPHYLLACEÆ.

DIANTHUS, L. PINK.

D. DELTOIDES, L. MAIDEN PINK.

Well established in field, Bar Harbor (Mary Minot); — field, Northeast Harbor (B. E. J. Gresham). Adventive from Europe.

SAPONARIA, L.

S. VACCARIA, L. COW HERB.

Uncommon. By roadside, Town Hill; in old grain field, Southwest Harbor (Rand). Adventive from Europe.

SILENE, L. CATCHFLY. CAMPION.

S. NOCTIFLORA, L. NIGHT-FLOWERING CATCHFLY.

Occasional in waste places. Roadside near Sargent Cove; Somesville ; Southwest Harbor (Rand). Adventive from Europe.

S. ARMERIA, L. SWEET WILLIAM CATCHFLY.

Occasionally escaped from gardens to roadsides and waste places. Southwest Harbor; Sea Wall; between Fernald and Norwood Coves (Rand). Adventive from Europe.

S. CUCUBALUS, Wibel. *S. inflata,* Smith. BLADDER CAMPION.

Well established in field, Bar Harbor (Mary Minot). Adventive from Europe.

S. NUTANS, L.

Well established in field, Bar Harbor, although not abundant (Mary Minot). Adventive from Europe.

LYCHNIS, L. COCKLE.

L. VESPERTINA, Sibth. EVENING LYCHNIS. WHITE CAMPION.

Rare. Waste ground, Fernald Point (Rand). Adventive from Europe.

L. GITHAGO (L.), Scop. CORN COCKLE.

Fields and roadsides; occasional. Southwest Harbor; Northeast Harbor (Rand); — Bracy Cove (Wm. C. Lane); — Seal Harbor (Redfield). Adventive from Europe.

ARENARIA, L. SANDWORT.

A. Grœnlandica (Retz), Spreng.

Frequent on mountain summits; often on rocky hills of less altitude; and less frequently on headlands and rocky shores by the sea, as at Bar Harbor (Rand). At Mt. Desert this plant blooms throughout the entire season from early June to October, the later flowers, however, being much smaller in size and fewer in number.

A. lateriflora, L.

Common; fields, thickets, and banks by the seashore. Also Duck Islands (Elizabeth G. Britton).

STELLARIA, L. CHICKWEED. STARWORT.

S. MEDIA (L.), Smith. COMMON CHICKWEED.

Common in cultivated and waste grounds. Naturalized from Europe.

S. longifolia, Muhl. LONG-LEAVED STARWORT.

Rare. Damp ground on Intervale Brook, near Hulls Cove (R. & R.).

S. GRAMINEA, L. ENGLISH STARWORT.

Becoming frequent in grassy places. Bass Harbor road, near Southwest Harbor; Southwest Harbor; Northwest Cove; Town Hill (Rand); — Northeast Harbor (Redfield). Adventive from Europe.

S. uliginosa, Murr. SWAMP STARWORT.

Rare. In wet ground and roadside ditch, east side of Northeast Harbor (T. Meehan & Redfield).

S. borealis, Bigel. NORTHERN STARWORT.

Frequent in wet ground. Little Harbor Brook Notch, Somesville (Rand); — Long Pond meadows (Wm. C. Lane); — Green Mt.; Bear Island (Redfield); — Otter Creek (T. G. White).

S. humifusa, Rottb.

Rare. Salt marsh, Little Cranberry Isle (Redfield). There is only one other station for this plant within the limits of Gray's Manual, but it is common farther north.

CERASTIUM, L. MOUSE-EAR CHICKWEED.

C. VULGATUM, L. Gray, Manual, 6th ed. *C. viscosum* of Man., 5th ed.

Fields and waste places; common. Naturalized from Europe.

C. arvense, L. FIELD CHICKWEED.

Rare. Duck Islands (Rand, Redfield, Annie S. Downs); — field near Ship Harbor (Redfield & Faxon).

SAGINA, L. PEARLWORT.

S. procumbens, L.

Springy places and wet rocks; frequent. Sea Wall (H. C. Jones); — Flying Mt.; Somesville ; Southwest Valley road, etc. (Rand); — Cranberry Isles (Redfield, Wm. C. Lane); — Duck Islands (Redfield); — Great Head (F. M. Day).

S. nodosa (L.), Fenzl. KNOTTY PEARLWORT.

Crevices of rocks and gravelly banks; rare. Bar Harbor (Rand, Kate Furbish); — The Cliffs, Seal Harbor (Rand).

BUDA, Adans. (*Spergularia*, Presl.) SAND SPURREY.

B. rubra (L.), Dumort. *Spergularia rubra* (L.), Presl. PINK SAND SPURREY.

Frequent; dry sandy soil, and occasionally on sea beaches. Manchester Point, Northeast Harbor; Bar Harbor (Rand); — Seal Harbor; Northeast Harbor ; Little Cranberry Isle (Redfield); — Great Cranberry Isle; Fernald Point (R. & R.); — on beach, Great Cranberry Isle (Rand). This plant in dry soil is strictly procumbent, usually undersized and dwarfed; on beaches it becomes large and widely spreading.

B. marina (L.), Dumort. *Spergularia salina*, Presl.

Sea beaches; rare. Sutton Island; Great Cranberry Isle (Rand).

B. borealis, S. Watson. *Spergularia borealis* (S. Watson), Robinson.

Frequent on sea beaches and salt marshes. Wasgatt Cove (J. L. Wakefield); — Somesville; Little Cranberry Isle (Redfield); — Southwest Harbor; Sea Wall; Norwood Cove; High Head; Bar Harbor, etc. (Rand); — Great Cranberry Isle (R. & R.).

SPERGULA, L. Spurrey.

S. ARVENSIS, L. CORN SPURREY.

A common weed in cultivated ground. Adventive from Europe.

Forma RUBRA.

Flowers deep pink. Field, Salisbury Cove (Rand).

PORTULACACEÆ. Purslane Family.

MONTIA, L. Blinks.

Sepals 2, ovate, persistent, herbaceous. Petals 5, united at base, 3 somewhat smaller. Stamens 3, rarely more, on the tube of the corolla. Ovary free, 3-ovuled: style 3-cleft, very short. Capsule 3-valved, 3-seeded. Seeds black, dull, tuberculate, rarely smoothish or shining.— A small branching glabrous succulent annual, with opposite leaves, and small axillary or racemose flowers. Bot. Cal., i. 77.

M. fontana, L.

Stems procumbent or ascending, 1 to 3 inches long: leaves spatulate to linear oblanceolate, 3 to 9 lines long: flowers a line long or less: capsule globose. (Bot. Cal., *l. c.*) Rare. Damp, brackish ground, Great Cranberry Isle (Rand); — Great Duck Island (Redfield). The only stations thus far known within the limits of Gray's Manual, or in Eastern U. S. Common, however, farther north, and on Pacific shores.

PORTULACA, L. Purslane.

P. OLERACEA, L. PURSLANE. PURSLEY.

Cultivated and waste grounds. Bar Harbor (Rand, W. H. Manning); — Long Pond (Rand). As yet a very uncommon weed on the Island, and probably introduced since 1880. Naturalized from Europe.

ELATINACEÆ. Waterwort Family.

ELATINE, L. Waterwort.

E. Americana (Pursh), Arn.

Rare. Margin of Somes Stream and of Mill Pond, Somesville (Rand, M. L. Fernald).

HYPERICACEÆ. St. John's-wort Family.

HYPERICUM, L. St. John's-wort.

* H. adpressum, Bart.

Rare. Adventive from farther south. Along the railroad, Green Mountain (Arnold Greene).

H. ellipticum, Hook.

Wet places and bogs; frequent. Pond Heath; Northeast Harbor; Southwest Harbor; High Head, etc. (Rand); — Somesville (R. & R.); — Squid Cove (Redfield).

H. PERFORATUM, L. Common St. John's-wort.

Fields and roadsides; frequent in some parts of the Island. A well known weed. Naturalized from Europe.

H. mutilum, L.

Frequent in low grounds and wet places. Outlet of Great Pond; Southwest Harbor; Somesville, etc. (Rand); — shores of Great Pond (Redfield).

H. Canadense, L.

Common in wet, sandy soil.

Var. majus, Gray.

Frequent. Southwest Harbor; Northeast Harbor; Denning Brook; Long Pond meadows; Somesville; shores of Jordan Pond and Great Pond; Bass Harbor, etc. (Rand); — Seal Harbor; Little Cranberry Isle (Redfield).

Two somewhat peculiar forms of this variety are sometimes met with on the Island: —

(*a*) Simple ; leaves ascending and somewhat appressed. Southwest Harbor, etc. (Rand).

(*β*) Leaves broadly lanceolate, more or less strongly five-nerved at the base. Intervale Brook; Southwest Harbor, etc. (Rand); — shores of Jordan Pond (Redfield).

H. nudicaule, Walt. *H. Sarothra,* Mx. Pine Weed. Orange Grass.

Sandy or gravelly soil, roadsides and mountain tops; common. The mountain form is exceedingly dwarfed, often being no more than a single unbranched stem, less than one inch in height.

ELODES, Adans. Marsh St. John's-wort.

E. campanulata (Walt.), Pursh. *E. Virginica* (L.), Nutt.

Common in swamps and bogs, and on borders of ponds.

MALVACEÆ. Mallow Family.

MALVA, L. Mallow.

M. rotundifolia, L. Common Mallow. Cheeses.

Waste and cultivated grounds about dwellings; not very common. Naturalized from Europe.

M. Alcea, L.

Escaped to roadsides between Town Hill and Salisbury Cove (M. L. Fernald). Adventive from Europe.

LINACEÆ. Flax Family.

LINUM, L. Flax.

L. usitatissimum, L. Common Flax.

Uncommon. Roadsides near Seal Cove, and about Southwest Harbor; abundant in grain field, Southwest Harbor (Rand). Adventive from Europe.

GERANIACEÆ. Geranium Family.

GERANIUM, L.

G. Robertianum, L. Herb Robert.

Frequent in damp, rocky places, especially at head of sea beaches. Abundant on Cranberry Isles (R. & R., F. M. Day); — Flying Mt. (Rand); — Bald Porcupine Island (W. H. Manning).

G. Carolinianum, L.

Frequent in waste places and clearings. Southwest Harbor (Harriet A. Hill); — roadside near Denning Pond (Annie S. Downs); — Somesville; Town Hill; Hulls Cove; High Head; Bubble Pond, etc. (Rand); — Seal Harbor (Redfield); — Bar Harbor (W. H. Manning).

OXALIS, L. Wood Sorrel.

O. Acetosella, L.

Common in mossy ground, deep cold woods. This plant is believed to be the true Irish Shamrock, although the emblem is now commonly represented by species of Trifolium.

O. corniculata, L., var. stricta (L.), Sav.

Common, mostly in open ground.

IMPATIENS, L. Balsam. Jewel Weed.

I. fulva, Nutt. Spotted Touch-me-not. Wild Balsam.

Moist places; common. A spurless form, Sea Wall (Rand).

Forma **albiflora.**

Flowers white or cream-color, spotted with pink; stems and foliage very pale. Southwest Harbor (Rand).

ILICINEÆ. Holly Family.

ILEX, L. Holly.

I. verticillata (L.), Gray. Black Alder.

Low grounds and thickets; common.

Var. tenuifolia, Torr.

Leaves petiolate, obovate, thin, smooth beneath except a slight pubescence on the midrib, uncinately serrate, obtuse, or more commonly mucronate-tipped; pistillate flowers 4–5-cleft, commonly solitary, short-pedicelled; berries scarlet. A shrub about 5° high with very slender branches. Torr. Fl. North. States, 338. A woodland form, appearing most distinct from the type.* On Denning Brook, Somesville (M. L. Fernald).

NEMOPANTHES, Raf. MOUNTAIN HOLLY.

N. fascicularis, Raf. *N. Canadensis* (Mx.), DC.

Damp ground; common.

VITACEÆ. VINE FAMILY.

AMPELOPSIS, Mx.

A. quinquefolia (L.), Mx. VIRGINIAN CREEPER. WOODBINE.

Common in cultivation, and often escaped to roadsides and waste places. Squid Cove; Southwest Harbor; Somesville (Rand). There is no satisfactory evidence that this plant is indigenous on the Island. Introduced from farther south.

SAPINDACEÆ. SOAPBERRY FAMILY.

ACER, L. MAPLE.

A. Pennsylvanicum, L. STRIPED MAPLE.

Common in woodlands.

A. spicatum, Lam. MOUNTAIN MAPLE.

Common in rocky woods.

A. saccharinum, Wang. SUGAR MAPLE.

Infrequent. Sargent District, etc. (Rand); — Seal Harbor, etc. (Redfield); — near Bar Harbor (W. H. Manning).

A. rubrum, L. RED MAPLE.

Common in swamps and damp ground.

* See also Britton in Bull. Torr. Bot. Club, xvii. 314.

ANACARDIACEÆ. Cashew Family.

RHUS, L. Sumach.

R. typhina, L. Staghorn Sumach.

Hillsides, etc.; frequent, but rather local.

R. Toxicodendron, L. Poison Ivy.

Thickets and low grounds; frequent, but local in its distribution. Common on cliffs and rocky banks by the seashore; — Pierce Head; Little Harbor; Hunters Beach; Roberts Point, etc. (R. & R.); — especially on the southern shore of the Island. It is also found in some abundance about Somesville and vicinity in low grounds and by roadsides. Poisonous to the touch.

POLYGALACEÆ. Milkwort Family.

POLYGALA, L. Milkwort.

P. paucifolia, Willd. Fringed Polygala.

Infrequent and local. Reported by various collectors from different parts of the region south and east of Salisbury Cove. Also found southwest of Youngs District (Clara L. Walley, Greenleaf, Lane & Rand).

P. sanguinea, L.

Infrequent. Fields, Bar Harbor; Southwest Harbor (Rand); — Long Pond meadows (Redfield); — "Mt. Desert" (R. H. Day).

P. verticillata, L.

Rare. Fields, Norwood Road, Southwest Harbor (Rand, Anna H. Bee).

LEGUMINOSÆ. Pulse Family.

TRIFOLIUM, L. Clover. Trefoil.

T. arvense, L. Rabbit-foot Clover.

Old fields, roadsides, etc.; common. Naturalized from Europe.

T. PRATENSE, L. RED CLOVER.

Fields and pastures; common. Naturalized from Europe.

T. repens, L. WHITE CLOVER.

Common everywhere in fields, pastures, and by waysides. Naturalized from Europe, and possibly indigenous northward.

T. HYBRIDUM, L. ALSIKE CLOVER.

Roadsides and fields; becoming common. This beautiful clover was rare on the Island ten or twelve years ago. Since that time it has appeared in increasing abundance every year. It does not seem, however, to be cultivated, or introduced intentionally. Naturalized from Europe.

T. AGRARIUM, L. HOP CLOVER.

Infrequent in fields and by roadsides. Southwest Harbor (Rand, Harriet A. Hill); — Seal Harbor (Redfield); — Bar Harbor (Mary Minot); — Beech Hill (Rand). Adventive from Europe.

T. PROCUMBENS, L. LOW HOP CLOVER.

Common in fields and by roadsides. Naturalized from Europe, but appearing indigenous. Small, simple, erect forms are common. This is "the real Irish Shamrock" of the newspapers, a long account of which appears regularly every two or three years. The error has been exposed so many times that it seems almost needless to refer to it here. (See *Oxalis Acetosella*, page 88.)

MELILOTUS, Juss. MELILOT.

M. OFFICINALIS (L.), Willd.

Sparingly introduced in grass fields. Seal Harbor (Lizzie Churchill). Adventive from Europe.

M. ALBA, Lam.

By waysides and in waste ground; more common than the last. Goose Cove (Wm. C. Lane); — Bar Harbor (F. M.

Day); — Somesville (Annie S. Downs); — Seal Harbor (Red field, Lizzie Churchill); — Eden; Fernald Point (Rand). Adventive from Europe.

MEDICAGO, L. MEDICK.

M. LUPULINA, L. BLACK MEDICK. SNAILS.

Sparingly introduced. Beach, Sea Wall (Rand); — Bar Harbor (Mary Minot). Adventive from Europe.

ROBINIA, L. LOCUST-TREE.

R. Pseudacacia, L. COMMON LOCUST.

Escaped from cultivation to roadsides. Southwest Harbor; Somesville; Town Hill, etc. (Rand); — Seal Harbor (Redfield). Adventive from the Middle States.

R. viscosa, Vent. CLAMMY LOCUST.

Escaped from cultivation. Roadside, Northeast Harbor (Redfield). Adventive from the mountains of the Southern States.

DESMODIUM, Desv. TICK TREFOIL.

D. acuminatum (Mx.), DC.

Rare. Clearing, Northwest Arm woods, Great Pond (Annie S. Downs). Perhaps introduced, as this species has not as yet been found elsewhere on the island.

VICIA, L. VETCH. TARE.

V. SATIVA, L. COMMON VETCH, or TARE.

Common, especially on sea beaches. Naturalized from Europe. A very pubescent form with flowers often peduncled, Bar Harbor (Mary Minot).

V. Cracca, L.

Frequent in fields. Southwest Harbor; Somesville, etc. (Rand); — Northeast Harbor (R. & R.); — Seal Harbor (Sara E. Boggs, Redfield); — Town Hill (Faxon & Rand); — Bar Harbor (Mary Minot, W. H. Manning).

LATHYRUS, L. EVERLASTING PEA.

L. maritimus (L.), Bigel. BEACH PEA.

Very common on sea beaches.

L. palustris, L. MARSH PEA.

Moist places near the sea; frequent. Aunt Mollys Beach; Southwest Harbor; Sea Wall; Great Cranberry Isle (Rand); — Bar Island, Bar Harbor (F. M. Day); — Little Cranberry Isle; Seal Harbor (Redfield); — Norwood Cove (M. L. Fernald).

L. PRATENSIS, L. FIELD PEA.

Well established in field, Bar Harbor (Mary Minot). Naturalized from Europe.

AMPHICARPÆA, Ell. HOG PEANUT.

A. monoica (L.), Ell.

Damp thickets on Somes Stream (Rand). Apparently indigenous, but perhaps introduced.

ROSACEÆ. ROSE FAMILY.

PRUNUS, L. PLUM. CHERRY.

P. Pennsylvanica, L. f. WILD RED CHERRY.

Rocky soil, woods and thickets; very common.

P. Virginiana, L. CHOKE CHERRY.

Waysides and thickets ; frequent. Salisbury Cove, etc. (Rand); — Somesville (R. & R.); — Squid Cove (Redfield); — Bar Harbor, etc. (F. M. Day).

P. serotina, Ehrh. WILD BLACK CHERRY.

Not uncommon about Somesville (Henry C. Jones, and others); — Bar Harbor (W. H. Manning). Blooming later than the preceding, which it much resembles.

SPIRÆA, L. MEADOW SWEET.

S. salicifolia, L. COMMON MEADOW SWEET.

Low grounds and damp hillsides; common.

S. tomentosa, L. HARDHACK.

Low grounds; common.

RUBUS, L. RASPBERRY. BLACKBERRY.

R. odoratus, L. PURPLE FLOWERING RASPBERRY.

Occasional by roadsides. Emery District; Southwest Harbor
(Rand); — Hulls Cove (R. H. Day). An evident escape from
cultivation. Adventive from beyond our limits.

R. Chamæmorus, L. BAKED APPLE BERRY.

Rare. The Heath, Great Cranberry Isle (R. & R.). Said to
grow in great abundance near Prospect Harbor, Gouldsborough,
on the mainland.

R. triflorus, Richards. WOOD RASPBERRY.

Common in damp woods and in swamps.

R. strigosus, Mx. WILD RED RASPBERRY.

Very common everywhere, especially in clearings and old
fields.

R. villosus, Ait. HIGH BLACKBERRY.

Waysides, fields, and thickets; very common.

Var. **frondosus** (Bigel.), Torr.

Frequent. Northwest Cove; about Somesville and elsewhere
(E. Faxon, R. & R.); — Bar Harbor (W. H. Manning).

Var. **Randii, Bailey.**

Low and diffuse, $1°-2\frac{1}{2}°$ high, the canes bearing very few and
weak prickles, or often entirely unarmed, very slender and soft,
sometimes appearing as if nearly herbaceous; leaves very thin
and nearly or quite smooth beneath and on the petioles, the
teeth rather coarse and unequal; cluster stout, with one or two
simple leaves in its base, not villous, and very slightly if at all

pubescent; flowers half or less the size of those of *R. villosus* ;
fruit small, dry, and "seedy." Woods, Southwest Valley road
(Rand).

R. Canadensis, L. Low Blackberry. Dewberry.

Dry fields and roadsides; frequent.

R. hispidus, L. Running Swamp Blackberry.

Low grounds and by waysides; common.

R. setosus, Bigel.

Stouter than *R. hispidus*, larger leaved, suberect or ascending, the older wood most densely clothed with slender, stiff,
slightly reflexed bristles; not evergreen; flowers usually small;
fruit reddish black, about 3″ high; leaflets mostly acute, or
short acuminate, generally 5 on the leaves of the sterile shoots,
and 3 on the flowering branches, short petiolulate or sessile;
pedicels and petioles often with a few weak bristles, pubescent.
(See N. L. Britton in Bull. Torr. Bot. Club, xx. 278, whence
the above description is mainly taken.) Not uncommon. Somesville; Beech Cliff; Oak Hill (Rand).

DALIBARDA, L.
D. repens, L.

Woods; common. Fertile flowers mainly, if not entirely,
cleistogamous, appearing rather earlier than the more showy
flowers. (See T. Meehan, Proceedings Acad. Nat. Sciences of
Phila., 1892, p. 371.)

GEUM, L. Avens.
G. album, Gmelin.

Thickets; rare. Wasgatt Cove (Wm. H. Dunbar); — Somesville (Rand); — near Somesville (Arnold Greene).

G. strictum, Ait.

Rare. Somesville (Rand). Perhaps introduced.

G. rivale, L. Water Avens.

Common in wet fields and meadows in the north and west of
the Island. Also Long Pond meadows (Redfield); — meadow at
Schooner Head (Robert B. Worthington); — Cold Brook (Rand).

FRAGARIA, L. STRAWBERRY.

F. Virginiana, Mill.

Very common everywhere.

* F. vesca, L.

Fields and rocky places; rare.

POTENTILLA, L. CINQUEFOIL.

P. Norvegica, L.

Fields and waste places; frequent.

P. Pennsylvanica, L.

Rare.   North of the Island (Wm. C. Lane?).

P. argentea, L. SILVERY CINQUEFOIL.

Dry ground; common.

P. palustris (L.), Scop. MARSH POTENTILLA.

Bogs; rare.   Somesville (Rand); — Great Cranberry Isle (Redfield).

P. fruticosa, L. SHRUBBY CINQUEFOIL.

Infrequent.   Asticou (Wm. C. Lane); — Sargent Mt. (Rand); — Jordan Mt. (Arthur Chase); — Southwest Harbor (Annie S. Downs); — Long Pond meadows (Redfield).   Generally in dry ground.

P. tridentata, Ait. WHITE POTENTILLA.

Shores and mountain summits; common.

P. Anserina. SILVER WEED.

Salt marshes and muddy beaches; common on the coast.

P. Canadensis, L. COMMON CINQUEFOIL. FIVE-FINGER.

Fields and waysides; common.   The typical form — which is low or dwarf, silky-hairy, with prostrate and decumbent stems — is rare.   Northeast Harbor (Rand).   The common form of the Island is var. *simplex* (Mx.), T. & G.   This is less hairy and greener, larger, the ascending stem 1°–2° long, seldom if ever creeping; from a thicker and harder caudex; leaflets obovate-oblong, sometimes almost glabrous. (Torr. & Gray, Fl. N. A., i. 443; Gray, Man., 5th ed. 154.) Although intermediate forms between this and the type are found, it seems that this is a good variety. Observations at Mt. Desert and elsewhere do not accord with those of Dr. N. L. Britton (Bull. Torr. Bot. Club, xviii. 365).

AGRIMONIA, L. AGRIMONY.

A. Eupatoria, L. COMMON AGRIMONY.

Rare. Woods, Roberts Point, Northeast Harbor; woods, Hadlock Upper Pond (Rand); — near Bar Harbor (W. H. Manning).

ROSA, L. ROSE.

R. Carolina, L. SWAMP ROSE.

Swamps and borders of streams; frequent.

R. lucida, Ehrh. COMMON WILD ROSE.

Abundant everywhere, usually in dry ground. A form with downy petioles, in rich soil, north of Long Pond (Redfield).

R. humilis, Marsh.

A plant answering to the description of this species has been found in woods, Somesville (M. L. Fernald). It is very desirable that the occurrence of this species should be further verified.

R. nitida, Willd. EARLY SWAMP ROSE.

In bogs throughout the Island, and on Cranberry Isles; common.

R. RUBIGINOSA, L. SWEET BRIER.

Rare. Naturalized from Europe in fields remote from dwellings, High Head; Bass Harbor (Annie S. Downs); — Seal Harbor (Rand).

R. CINNAMOMEA, L. CINNAMON ROSE.

Stems 5°–8° high with brownish-red bark, and some straightish prickles; leaves pale, downy beneath; flowers small, pale

7

pink, cinnamon-scented, mostly double, not showy.   Gray,
Field, For. & Gard. Bot., 127.   Roadsides, escaped from
gardens.   Hulls Cove; Oak Hill (Rand).   Introduced from
Europe.

PYRUS, L. APPLE. PEAR.

P. MALUS, L. APPLE.

Infrequently spontaneous by waysides, in old fields, etc.
Northeast Harbor ; Somesville ; Canada Valley ; Sutton Island
(Rand).

P. arbutifolia (L.), L. f., var. **melanocarpa** (Mx.), Hook.
BLACK CHOKEBERRY.

Common in both wet and dry ground.   Very variable in height
from 4°-5° in swamps to 6'-1° on mountains.   A double-
flowered form, Breakneck Ponds (Rand).

Forma **pubescens.**

Pedicels and petioles very tomentose.   Not uncommon.
Somesville (R. & R., M. L. Fernald); — Bar Harbor (Mary
Minot).

P. Americana (Marsh.), DC.   MOUNTAIN ASH.

Rocky woods; common.   Leaves commonly less taper-pointed,
and a darker green in color than farther south.   Somewhat approaching the next in general appearance.

P. sambucifolia, Cham. & Schlecht.   NORTHERN MOUNTAIN ASH.

Rare.   Beech Cliff (E. Faxon); — Southwest Harbor (M. L.
Fernald; — Beech Hill (Rand).

CRATÆGUS, L. HAWTHORN. THORN.

C. coccinea, L.   SCARLET-FRUITED THORN.

Infrequent.   Little Harbor Brook Notch (Rand); — Denning
Brook (M. L. Fernald).

Var. **macracantha** (Lodd.), Dudley.

Frequent by waysides, rocky banks, beaches, etc.   The common Thorn of the Island.

AMELANCHIER, Medic. SHADBUSH. SUGAR PEAR.

A. Canadensis (L.), Medic. SHADBUSH.

Rocky woods; common. Large trees of this species in Little Harbor Brook Notch.

Var. **oblongifolia,** T. & G.

Common.

SAXIFRAGACEÆ. SAXIFRAGE FAMILY.

SAXIFRAGA, L. SAXIFRAGE.

S. Virginiensis, Mx. EARLY SAXIFRAGE.

Rocky places; infrequent. Valley Cove; Dog Mt.; Flying Mt.; Beech Cliff (Rand); — Sargent Mt. (Greenleaf, Lane & Rand); — Schooner Head (Clara L. Walley); — ledges on road between Seal Harbor and Hunters Brook (Redfield).

MITELLA, L. MITREWORT.

M. nuda, L.

Cool, mossy woods; rare. Hadlock Valley (Redfield); — Cold Brook (R. & R.); — woods, head of Barcelona meadow (Rand).

CHRYSOSPLENIUM, L. GOLDEN SAXIFRAGE.

C. Americanum, Schwein.

Brooks and wet places; infrequent. On trail between Jordan Pond and Northeast Harbor; Doctors Brook; Little Harbor Brook; Canada Brook; Cold Brook (Rand); — Two Mile Brook (M. L. Fernald).

RIBES, L. GOOSEBERRY. CURRANT.

R. oxyacanthoides, L. WILD GOOSEBERRY.

Common, usually in rocky ground.

R. lacustre, Poir.

Rare. In wet pasture, Great Cranberry Isle (R. & R.).

R. prostratum, L'Hér. Skunk Currant.

Common in rocky places.

R. floridum, L'Hér. Wild Black Currant.

Uncommon. Clearing, Canada Valley (Rand); — on Somes Stream (R. & R., M. L. Fernald); — Beech Hill (Redfield). Without doubt introduced in the last-named station, and doubtless escaped from cultivation in the others.

R. rubrum, L. Red Currant.

Sparingly escaped from cultivation. Beech Hill, etc.; near High Head, remote from dwellings (Rand).

CRASSULACEÆ. Orpine Family.

SEDUM, L. Stonecrop. Orpine.

S. acre, L. Mossy Stonecrop.

Sparingly escaped from cultivation to roadsides, rocky places, etc. Roadside, Southwest Harbor (William H. Dunbar); — established in abundance on rocks, Southwest Harbor (Henry L. Rand) ; — among stones, near the cemetery, Somesville (Redfield). Naturalized from Europe.

S. Telephium, L. Live-for-ever.

Escaped from cultivation to roadsides and fields; frequent. Southwest Harbor; Fernald Cove ; Great Cranberry Isle (Rand); — Hulls Cove (F. M. Day); — Somesville (R. & R.). Adventive from Europe.

S. Rhodiola, DC. Roseroot.

Rare. Dog Mt. (Henry C. Jones, Rand); — Egg Rock (Henry Smith).

DROSERACEÆ. Sundew Family.

DROSERA, L. Sundew.

D. rotundifolia, L. Round-leaved Sundew.

Common in sphagnous bogs, and in wet places generally.

D. intermedia, Drev. & Hayne, var. Americana (Willd.), DC.
Bogs and borders of ponds; common.

HAMAMELIDEÆ. Witch Hazel Family.

HAMAMELIS, L. Witch Hazel.

H. Virginiana, L.
Occasional in woods and by roadsides.

HALORAGEÆ. Water Milfoil Family.

MYRIOPHYLLUM, L. Water Milfoil.

M. verticillatum, L.
Rare. Ripples Pond (M. L. Fernald, Rand).

PROSERPINACA, L. Mermaid Weed.

P. palustris, L.
Rare. Meadow at head of Northeast Creek (Rand, M. L. Fernald).

HIPPURIS, L. Mare's Tail.

H. vulgaris, L.
Uncommon. Shallow pools at Sea Wall and vicinity, near shore (Henry C. Jones, Elizabeth G. Britton, Rand); — marsh, Great Duck Island (Redfield); — not rare, Great Cranberry Isle (Rand, Redfield, Arnold Greene).

CALLITRICHE, L. Water Starwort.

C. verna, L.
Frequent in muddy streams and ditches. Ditch, Bass Harbor (Rand); — Somes Stream (R. & R.); — Hunters Brook (Redfield) ; — Two Mile Brook ; Northeast Creek (M. L. Fernald).

MELASTOMACEÆ. Melastoma Family.

RHEXIA, L. Meadow Beauty.

R. Virginica, L.

Sandy or gravelly pond shores; infrequent. Duck Brook meadows; Breakneck Ponds; Great Pond (Rand); — Eagle Lake; Witch Hole (Redfield).

LYTHRACEÆ. Loosestrife Family.

DECODON, Gmelin. Swamp Loosestrife.

D. verticillatus (L.), Ell. *Nesæa verticillata* (L.), HBK.

Rare. Swamp, Somes Pond (R. & R.).

ONAGRACEÆ. Evening Primrose Family.

LUDWIGIA, L. False Loosestrife.

L. palustris (L.), Ell. Water Purslane.

Abundant on flats, Ripples Pond (Rand). The only station thus far reported on the Island.

EPILOBIUM, L. Willow Herb.

E. angustifolium, L. Fireweed.

Common, especially in clearings and in burnt ground.

Forma **albiflorum.**

Flowers pure white. Little Cranberry Isle (William H. Dunbar).

E. lineare, Muhl.

Bogs and in wet ground; common. A simple form approaching *E. palustre*, Western Mt. (Rand).

Forma **latifolium.**

Leaves larger, often 4″ broad and 2′ long. Bog, north of Beech Hill (Rand).

E. strictum, Muhl.

Bogs and wet ground; frequent. Near Sea Wall; High Head; Southwest Harbor; Great Cranberry Isle (Rand); — Salisbury Cove (Clara L. Walley). A form said by Prof. Trelease to be "perhaps crossed with *E. lineare,*" bog, north of Beech Hill (Rand).

E. coloratum, Muhl.

Apparently rare on the Island. Bog, north of Beech Hill; Southwest Harbor (Rand).

A hybrid, *E. coloratum* × *E. adenocaulon,* roadside, south of Bubble Pond (Rand).

E. adenocaulon, Hausskn.

Common in low ground.

ŒNOTHERA, L. EVENING PRIMROSE.

Œ. biennis, L. COMMON EVENING PRIMROSE.

Fields, waysides, and sea beaches; common. Variable. A very pubescent form with long white hairs, Hadlock farm, near Seal Harbor (Redfield); — Seal Harbor (Sara E. Boggs).

Œ. pumila, L.

Common in dry soil. Often much dwarfed.

CIRCÆA, L. ENCHANTER'S NIGHTSHADE.

C. alpina, L.

Common in damp, shady woods.

CUCURBITACEÆ. GOURD FAMILY.

ECHINOCYSTIS, T. & G. WILD BALSAM APPLE.

E. lobata (Mx.), T. & G.

Extensively cultivated throughout the Island, and often spontaneous and persistent in waste places and by waysides. Bar Harbor; Southwest Harbor, etc. (R. & R.) Adventive from the West.

FICOIDEÆ.

MOLLUGO, L. INDIAN CHICKWEED.

M. VERTICILLATA, L. CARPET WEED.

Rare. Bar Harbor (W. H. Manning). A very lately introduced weed on the Island. Probably naturalized from Tropical America.

UMBELLIFERÆ. PARSLEY FAMILY.

DAUCUS, L. CARROT.

D. CAROTA, L.

Old fields; occasional. Long Pond (Redfield); — Sea Wall (Rand); — Bar Harbor (Mary Minot).

CONIOSELINUM, Fisch. HEMLOCK PARSLEY.

C. Canadense (Mx.), T. & G.

Wet woods and meadows; frequent. Seal Harbor; Long Pond meadows (Redfield); — Hadlock Upper Pond; Little Harbor Brook Valley, etc. (Rand).

HERACLEUM, L. COW PARSNIP.

H. lanatum, Mx.

Frequent, especially near sea beaches. Somesville (R. & R.); — Fernald Point; Southwest Harbor; Sea Wall; Great Cranberry Isle, etc. (Rand); — Baker Island (Redfield).

PASTINACA, L. PARSNIP.

P. SATIVA, L.

Roadsides, waste grounds, and old fields; frequent. Northeast Harbor (Henry C. Jones); — Long Pond; Somesville; Southwest Harbor; Bar Harbor; Canada Valley, etc. (Rand). Naturalized from Europe.

LIGUSTICUM, L. LOVAGE.

L. Scoticum, L. SCOTCH LOVAGE.

Sea beaches, banks and rocks near salt water; common.

CŒLOPLEURUM, Ledeb.

C. Gmelini (DC.), Ledeb. *Archangelica Gmelini*, DC.

Frequent on the seashore in damp ground. Plant very strongly aromatic.

SIUM, L. WATER PARSNIP.

S. cicutæfolium, Gmelin.

Brooks and pond shores; common.

CARUM, L. CARAWAY.

C. CARUI, L.

Common in fields and waste places about settlements. Naturalized from Europe.

CICUTA, L. WATER HEMLOCK.

C. maculata, L.

Wet grounds; common. Roots poisonous.

C. bulbifera, L.

Wet places ; infrequent. Somesville ; Northeast Harbor (Rand). Also poisonous.

HYDROCOTYLE, L. WATER PENNYWORT.

H. Americana, L.

Wet places, — woods and meadows; frequent. Hadlock Upper Pond; High Head; Southwest Valley road ; Town Hill, etc. (Rand); — Somesville (R. & R., M. L. Fernald). Stems stoloniferous, especially in late summer.

SANICULA, L. Black Snakeroot.

S. Marylandica, L.

Wet woods and meadows; infrequent. Little Harbor Brook Notch (R. & R., Arnold Greene); — Cold Brook; Long Pond meadows (Rand).

ARALIACEÆ. Ginseng Family.

ARALIA, L. Wild Sarsaparilla.

A. racemosa, L. Spikenard.

Woods; infrequent. Little Harbor Brook Notch (Rand, Redfield); — Intervale Brook (F. M. Day); — Hadlock Valley (G. Hunt); — Southwest Valley road; Wild Cat Valley (Rand); — roadside east of Seal Harbor ; Bubble Pond (Redfield).

A. hispida, Vent. Bristly Sarsaparilla.

Open rocky places and burnt clearings; common.

A. nudicaulis, L. Wild Sarsaparilla.

Rich rocky woods; common.

CORNACEÆ. Dogwood Family.

CORNUS, L. Cornel. Dogwood.

C. Canadensis, L. Bunchberry. Dwarf Cornel.

Very common; woods and everywhere. Often blooming until late fall.

C. circinata, L'Hér. Round-leaved Cornel.

Woods and thickets; infrequent. Echo Notch (R. & R.); — Little Harbor Brook Notch; High Head; Northwest Cove; Somesville (Rand).

C. alternifolia, L. f. Alternate-leaved Cornel.

Woods and copses; common.

Division II. GAMOPETALÆ.

CAPRIFOLIACEÆ. Honeysuckle Family.

SAMBUCUS, L. Elder.

S. Canadensis, L. Common Elder.

Frequent in rich soil, but nowhere very abundant. Somesville; Northeast Harbor; Southwest Harbor; Gilmore Meadow, etc. (Rand); — Long Pond meadows, etc. (Redfield); — Bar Harbor (F. M. Day).

S. racemosa, L. *S. pubens*, Mx. Red-berried Elder.

Rocky places and waysides; more common than the last. A form with yellowish colored fruit, near Northeast Harbor (Theodore G. White).

VIBURNUM, L. Arrow Wood.

V. lantanoides, Mx. Hobble Bush.

Common; rocky woods, — especially on mountain brooks, — and sometimes in low ground.

V. acerifolium, L. Maple-leaved Viburnum.

Thickets and borders of woods; frequent; rare in the southeastern part of the Island.

V. dentatum, L. Arrow Wood.

Rare. Meadow at head of Northeast Creek (Rand, Redfield, M. L. Fernald).

V. cassinoides, L. Withe-rod.

Rocky woods, moist banks, and wet places; common. Also on mountain summits.

LINNÆA, Gronov. Twin Flower.

L. borealis, Gronov. Twin Flower.

Woods, especially in sandy soil; common. Also Cranberry Isles and Duck Islands (Redfield).

LONICERA, L. Honeysuckle.

L. ciliata, Muhl. Fly Honeysuckle.

Rocky woods; infrequent. Hadlock Upper Pond; Sargent Mt. Gorge; Little Harbor Brook Notch; Northwest Arm woods, etc. (Rand); — Squid Cove (Wm. C. Lane); — Hadlock Valley; Jordan Pond road; Bear Island (Redfield); — Norway Drive (Mary Minot); — Bald Porcupine Island (W. H. Manning).

L. cærulea, L. Mountain Fly Honeysuckle.

Damp ground; frequent.

DIERVILLA, Adans. Bush Honeysuckle.

D. trifida, Mœnch.

Common in rocky ground; woods and clearings.

RUBIACEÆ. Madder Family.

HOUSTONIA, L.

H. cærulea, L. Innocents. Bluets. Quaker Ladies.

Common in moist grassy places.

MITCHELLA, L. Partridge Berry.

M. repens, L.

Thickets and woods, especially under Coniferæ; frequent.

GALIUM, L. Bedstraw.

G. verum, L. Yellow Bedstraw.

Well established for years in fields, Bar Harbor (Mary Minot, M. L. Fernald). Adventive from Europe.

G. Mollugo, L.

Established in field, Bar Harbor (Mary Minot). Adventive from Europe. A somewhat pubescent form with revolute, sharply pointed leaves, Bar Harbor (Mary Minot).

G. Aparine, L. CLEAVERS.

Rare; yet in abundance on beach, Fish Point, Great Cranberry Isle (Rand). Probably introduced.

G. trifidum, L. SMALL BEDSTRAW.

Common in wet ground. Variable.

Var. pusillum, Gray.

Not uncommon in cold sphagnum bogs. Little Cranberry Isle; Great Duck Island (Redfield); — Southwest Harbor (M. L. Fernald).

G. asprellum, Mx. ROUGH BEDSTRAW.

Roadsides and low thickets; frequent. Robinson Mt. (Wm. H. Dunbar); — Somesville; Emery District (R. & R.); — Sea Wall; Oak Hill, etc. (Rand).

G. triflorum, Mx. SWEET-SCENTED BEDSTRAW.

Woods; common.

COMPOSITÆ. COMPOSITE FAMILY.

EUPATORIUM, L. THOROUGHWORT.

E. purpureum, L. JOE-PYE WEED.

Low grounds, brooksides, and meadows; frequent.

E. perfoliatum, L. BONESET. THOROUGHWORT.

Low grounds and wet roadsides; common.

SOLIDAGO, L. GOLDEN ROD.

S. squarrosa, Muhl.

Rare; roadsides and thickets. Emery District (Rand, Annie S. Downs); — foot of Western Mt. (Rand); — "Salisbury Woods" (Clara L. Walley).

S. latifolia, L.

Damp, low woods, especially by brooksides; frequent. Head of Hadlock Upper Pond; foot of Western Mt.; Little Harbor

Brook Notch; Beech Mt. Notch; and elsewhere (Rand); — path to Bubble Pond ; path to Newport Pond ; Hadlock Valley (Redfield); — Dog Mt., in dry, open ground (Rand).

S. bicolor, L.   WHITE GOLDEN ROD.

Roadsides and fields; common.

Var. concolor, T. & G.

Roadside, south of High Head. The plant, however, is not a very well marked form of this variety.

S. Virgaurea, L.

Stem erect, sparingly branched, 4′–24′ high, glabrous or pubescent with curled hairs; leaves linear or lanceolate-oblong, 1′–4′ long, obscurely toothed, obtuse or acute; heads crowded, 4″ long, shortly peduncled, golden yellow; bracts of the involucre linear, acute, glabrous, green, margins scarious; ray flowers 10–12, spreading; disk flowers 10–20; achene pubescent, pappus white. Hooker, Fl. Brit. Isles, 205. (See also Bull. Torr. Bot. Club, xx. 207.)   Hio, Southwest Harbor; foot Pemetic Mt.; Great Cranberry Isle; path on Jordan Mt.; Frenchman Camp road (Rand); — Seal Harbor (R. & R.). This Island form is very like var. *angustifolia*, Gaud., and var. *ericetorum*, DC., of the Old World, with lower leaves oblong lanceolate, long petioled, and upper leaves narrower. There are other forms, with narrower, thinner leaves, approaching *S. humilis*, Pursh, but hardly to be placed under that species. Seal Harbor (Redfield); — foot of Western Mt.; Great Cranberry Isle; Dog Mt.; east peak of Western Mt.; Frenchman Camp road (Rand).

Var. Randii, Porter.

More or less glutinous; stems stout, erect, 1°–2° high, often dark purple, puberulent, or sometimes glabrate below; radical and lower leaves obovate or oblanceolate, acute, serrate, — cauline lanceolate or elliptical-lanceolate, sparingly serrate or entire, glabrous; inflorescence an ample branched panicle or loose virgate thyrse; heads 3″ or more long; outer scales of the involucre mostly ovate or lance-ovate and bluntish, sometimes almost linear and acute, inner ones oblong-linear, yellowish, with scarious margins and acute or acuminate tips; achenes pubescent or nearly smooth. Bull. Torr. Bot. Club, xx. 208. Abundant; dry fields, roadsides, and among rocks, especially in the southern part of the Island. Also found on the mountains, distinct, or imperceptibly passing into the next variety. Frenchman Camp; Seal Harbor (R. & R.); — Sea wall, Long Pond; Northeast Harbor; Southwest Harbor; The Cliffs, Seal Harbor; Hunters Beach Head; Long Pond meadows; Sargent Mt.; Dog Mt.; Pemetic Mt.; Western Mt.; Great Cranberry Isle, etc. (Rand).

Var. monticola, Porter. *S. puberula*, Nutt., var. *monticola*, Porter, Bull. Torr. Bot. Club, xix. 129.

Stems 3'–12' high, often slender; inflorescence a short, compact, or sometimes loose thyrse, 2'–4' long; heads 1½''–3'' long; scales of the involucre variable, ovate and bluntish or oblong and obtuse, inner ones not elongated. Bull. Torr. Bot. Club, xx. 209. Common on mountain summits, and occasionally at lower altitudes, even at the sea level. Sargent Mt.; Pemetic Mt. ; Jordan Mt.; Western Mt.; Dog Mt.; Hunters Beach Head; Seal Harbor; Great Cranberry Isle (Rand).

Var. Redfieldii, Porter.

Very glutinous; stems stout and rigid, 16'–18' high; leaves thickish or coriaceous; branches of the panicle starting from half-way down the stem or even from the base, strict, erect, bearing short clusters of heads in the upper bracts; heads small, 2''–3'' long; scales of the involucre short, more or less scarious. Its inflorescence is strikingly like that of *S. juncea*, Ait., var. *ramosa*, Porter & Britt. Bull. Torr. Bot. Club, xx. 209. Rare. Seal Harbor, etc. (Redfield); — foot of Western Mt.; Great Cranberry Isle (Rand).

S. sempervirens, L. SEA GOLDEN ROD.

Frequent on sea cliffs; common in salt or brackish marshes, and on muddy beaches.

S. puberula, Nutt.

Dry, open ground, and by waysides; common; — less frequent in woods. Very variable in form of inflorescence, etc. A form with panicle much-branched, branches erect, much resembling *S. juncea*, Ait., var. *ramosa*, Porter & Britt., in clearings, Sunken Heath, and elsewhere (Rand). A form with inflorescence axillary, much prolonged, Emery District, and elsewhere (Rand).

S. rugosa, Mill. *S. altissima*, T. & G., *non* L.

Fields, thickets, and roadsides; common.

S. neglecta, T. & G.　Swamp Golden Rod.

Swamps, bogs, and meadows. Especially common in sphagnum bogs.

Var. **linoides** (T. & G.), Gray.

Sphagnum bogs; infrequent. Great Heath; The Heath, Great Cranberry Isle (Rand).

S. juncea, Ait.　Early Golden Rod.

Dry ground; common. The earliest flowering species of the genus on the Island. A form more or less pubescent, Hio, Southwest Harbor (Rand). A form approaching var. *ramosa*, Porter & Britt., Jordan Mt. (Rand).

S. serotina, Ait.

Rare. Copses, Long Pond meadows (Redfield); — Salisbury Cove (Clara L. Walley).

Var. **gigantea** (Ait.), Gray.

Copses and low grounds; frequent. A low form, Long Pond meadows (Redfield). A form approaching *S. rupestris*, Raf., Long Pond meadows (Redfield).

S. Canadensis, L.

Roadsides, fields, and thickets; common.

Var. **glabrata**, Porter.

Low, slender 2°–3° high, stems glabrous or glabrate below, puberulent above; leaves numerous, crowded, linear-lanceolate, tapering into a long acumination, upper ones entire, lower ones with a few sharp serratures, scabrous on the veins beneath;

panicles small, with filiform branches; bracts acute or acutish. Roadsides and thickets; infrequent. Near Hadlock Brook, Wasgatt Cove (Rand);—road between Frenchman Camp and Hadlock farm; Frenchman Camp; above Long Pond meadows (Redfield).

S. nemoralis, Ait.

Dry grounds; very common. A form with axillary, much prolonged inflorescence, clusters distant,—between Southwest Harbor and Bass Harbor (Rand).

S. lanceolata, L.

Roadsides and fields; very common.

ASTER, L. ASTER.

A. macrophyllus, L. GREAT-LEAVED ASTER.

Woods and clearings; common. Flowering more abundantly when in clearings or in open ground. Often appearing with few or none of the characteristic root-leaves. To this form, it seems, should be referred specimens collected near Bubble Pond (Redfield), and named by Dr. Asa Gray *A. Herveyi.*

A. radula, Ait. ROUGH-LEAVED ASTER.

Low grounds and borders of swamps; common. A form approaching var. *strictus*, Gray, roadside between Town Hill and Emery District (Rand).

A. undulatus, L.

Rare. Frenchman Camp road (R. & R.).

A. cordifolius, L. HEART-LEAVED ASTER.

Wooded banks and waysides; rare. Somesville; Juniper Cove; near head of Northeast Harbor (Rand).

A. Lindleyanus, T. & G.

Dry ground; rare. Southwest Valley road; High Head (Rand);—Thompson Island (Annie S. Downs);—Frenchman Camp road (Redfield).

A. polyphyllns, Willd.

Infrequent and local; in greatest abundance on the mountains. Dog Mt.; Jordan Mt.; Pemetic Mt.; Long Pond meadows (Rand); — Frenchman Camp; on Hunters Brook, near French- man Camp (R. & R.). The forms from Jordan Mt. are much dwarfed.

A. ericoides, L.

Rare. Wayside, road to Jordan Pond (R. & R.).

A. vimineus, Lam.

Rare. Frenchman Camp (Rand).

A. diffusus, Ait.

Fields, thickets, and waysides; very common. Variable. A form with purple ray flowers and panicles less elongated, Little Harbor (Rand).

Var. thyrsoideus, Gray.

Rare. Seal Harbor (R. & R.).

A. Tradescanti, L.

Rare. Southern foot of Western Mt. (Rand).

A. paniculatus, Lam.

Moist ground; frequent and widely distributed, especially in the northern part of the Island. Shore of Northwest Arm, Great Pond; Liscomb Brook, and elsewhere in Emery Dis- trict; Oak Hill and northward; Somesville; Ripples Pond; Bass Harbor; road to Great Pond, Southwest Harbor, etc. (Rand).

A. salicifolius, Ait.

Low grounds and roadsides; frequent. Southwest Harbor; southern foot of Western Mt.; Somesville; Beech Hill; Oak Hill; Town Hill; Bass Harbor; Northwest Cove; Eden P. O., etc. (Rand).

A. junceus, Ait.

Rare. Roadside between Southwest Harbor and Bass Har- bor; between Town Hill and Emery District (Rand).

A. longifolius, Lam. Not of Gray, Manual, 5th ed.

Low grounds; infrequent. Roadside thicket near Juniper Cove; Cliff walk, Seal Harbor; Meadow Brook, Oak Hill; roadside between Town Hill and Emery District; Eden P. O.; Somesville (Rand).

A. Novi-Belgii, L.

Abundant everywhere in both wet and dry ground. Very variable in foliage, size of heads, color of ray flowers, etc. The common form has smooth, thick leaves. A form with very narrow leaves, Somesville; Bass Harbor; Emery Cove; foot of: Western Mt.; Pond Heath, etc. (Rand). Forms of this species apparently pass into *A. puniceus*, L., var. *lucidulus*, Gray.

Forma **albiflorus.**

Ray flowers pure white. Cliffs east of Seal Harbor (Rand).

Forma **roseus.**

A salt marsh form, low, only 1° high; stems dark purple; leaves linear, thick, with purple midrib, the lower stem leaves bearing abundant axillary clusters of small leaves; rays bright pink. Near Bass Harbor (Rand).

Var. litoreus, Gray.

Salt marshes or wet shores; infrequent or rare. Mouth of Northeast Creek; Somes Harbor; Great Cranberry Isle (Rand). Forms approaching this variety are abundant, but the variety itself is seldom found.

A. patulus, Lam.

Rare. Meadow, head of Northeast Creek (M. L. Fernald); — Town Hill road, Somesville (Rand).

A. tardiflorus, L.

Rare. Wood road to Broad Cove; road from Town Hill to Thomas Bay (Rand).

A. puniceus, L.

Common in wet ground.

Var. lucidulus, Gray.

Frequent in wet places and moist ground. Seal Harbor (Redfield); — Town Hill; Emery Cove; Youngs District; Eden P. O., etc. (Rand). Apparently more abundant in the northern part of the Island.

Var. lævicaulis, Gray.

Infrequent. Beech Mt. Notch (Rand); — stream south of Bubble Pond (R. & R.).

A. umbellatus, Mill. *Diplopappus umbellatus,* T. & G.

Roadsides, fields, wood clearings, and dry places ; very common.

A. acuminatus, Mx.

Woods and clearings; common.

A. nemoralis, Ait.

Peat bogs and open swamps; common. Also summit of Green Mt. in boggy depressions (Rand.)

Var. Blakei, Porter.

Stems $1°-2\frac{1}{2}°$ high, simple, or often branched, inclined to be flexuous; leaves not crowded as in the type, $2\frac{1}{2}'-3'$ long, $\frac{1}{2}'-\frac{3}{4}'$ broad, lanceolate, coarsely toothed or entire, margins not revolute, thinnish; heads few or several, sometimes solitary, showy; rays lilac-purple. — Intermediate between *A. nemoralis* and *A. acuminatus*, to both of which some of its forms make a near approach. North border of Somes Pond (Rand).

ERIGERON, L. Fleabane.

E. Canadensis, L. Horseweed. Butterweed.

A common weed by roadsides and in waste places, becoming yearly more abundant everywhere on the Island.

E. strigosus, Muhl. Smaller Daisy Fleabane.

Fields and waysides; common. Rays sometimes more or less deeply tinged with violet.

ANTENNARIA, Gærtn. EVERLASTING.

A. plantaginifolia (L.), Hook. MOUSE-EARS. LADIES' TOBACCO.
Dry soil; common.

ANAPHALIS, DC. PEARLY EVERLASTING.

A. margaritacea (L.), Benth. & Hook. *Antennaria marga-*
ritacea (L.), R. Br. PEARLY EVERLASTING.
Dry soil, clearings, etc.; common.

GNAPHALIUM, L. CUDWEED.

G. polycephalum, Mx. SWEET EVERLASTING.
Dry fields; common. Very sweet-scented.

G. decurrens, Ives.
Fields, in sandy soil; infrequent. Sawyer Cove; Southwest
Harbor; Seal Harbor (Rand); — Little Cranberry Isle (Redfield).

G. uliginosum, L. LOW CUDWEED.
Roadsides; common in damp soil. Also Green Mt. (Redfield).

AMBROSIA, L. RAGWEED.

A. artemisiæfolia, L. RAGWEED.
A common weed in waste places. Many people suppose its
pollen an effective cause of hay fever, and find relief on the
Island owing to the supposed absence of this plant. This
relief however, it seems, must be attributed to some other
cause, as the plant in question grows everywhere, — even on
the Cranberry Isles, — in more or less abundance, and is spreading with the increasing settlement of the Island.

RUDBECKIA, L. CONE FLOWER.

R. hirta, L. YELLOW DAISY. BLACK-EYED SUSAN.
Naturalized in grass fields from the West. Very common
about Somesville and Bar Harbor, and more or less abundant
all over the Island. A fastigiate form, Seal Harbor (Redfield).

HELIANTHUS, L.   Sunflower.

H. annuus, L.   Common Sunflower.

Waste places, Fernald Point; Bar Harbor (Rand); — Seal
Harbor (Redfield).   Escaped from cultivation.

BIDENS, L.   Bur-Marigold.

B. frondosa, L.   Beggar Ticks.   Devil's Pitchfork.

Low grounds, wet places, and damp roadsides; common, and
spreading.

B. cernua, L.   Smaller Bur-Marigold.

Wet places; frequent.   Sawyer Cove; Mill Cove; Norwood
Cove; Valley Cove; northern foot of Beech Hill; Southwest
Harbor, etc. (Rand); — Somesville (R. & R.); — Great Cran-
berry Isle (Redfield).

ANTHEMIS, L.   Chamomile.

A. Cotula, L.   *Maruta Cotula* (L.), DC.   Mayweed.

Common by roadsides and in waste places.   Naturalized
from Europe.

ACHILLEA, L.   Yarrow.

A. Millefolium, L.   Common Yarrow.

Dry soil; common in settlements, and often remote from
dwellings.   Naturalized from Europe, but also indigenous.

Forma **rosea.**

Ray flowers rose to deep rose-red in color.   Frequent.   Northeast Harbor: Fernald Point road; southern foot of Western
Mt.; Southwest Harbor, etc. (Rand); — Somesville (R. & R.).

A. Ptarmica, L.   Sneezewort.

Doubtless an escape, although established for years by roadside spring, far from dwellings, Southwest Harbor (Rand).
Adventive from Europe or farther north.

CHRYSANTHEMUM, L. Ox-eye Daisy.

C. Leucanthemum, L. Daisy. White-weed.

Fields and meadows; very common. Naturalized from Europe. A form with tubular ray flowers, Ovens (Rand).

TANACETUM, L. Tansy.

T. vulgare, L. Common Tansy. Gold Buttons.

Near old dwellings, and by waysides; frequent. Jordan Pond; Beech Hill; "Sound"; Great Cranberry Isle, etc. (Rand); — Bass Harbor Head; Somesville (R. & R.); — mouth of Duck Brook (R. H. Day). Naturalized from Europe.

ARTEMISIA, L. Wormwood.

A. vulgaris, L. Mugwort.

Occasional in waste places. West side of Southwest Harbor; Hulls Cove (Rand). Adventive from Europe.

A. Stelleriana, Besser. False Dusty Miller.

Rare. Beach, Mt. Desert Narrows (R. & R., Annie S. Downs). Adventive from the north.

PETASITES, Gærtn. Sweet Coltsfoot.

P. palmata (Ait.), Gray.

Wet ground; rare. Cold Brook (E. Faxon, R. & R.).

SENECIO, L. Groundsel.

S. vulgaris, L. Common Groundsel.

Waste grounds and sea beaches; frequent, especially near the seashore and on the islands. Also, Green Mt. (Annie S. Downs). On Flying Mt. is found a low, slender form with aromatic foliage (Rand). Naturalized from Europe.

S. aureus, L. Golden Ragwort.

Rare. Meadow, High Head (Rand).

ERECHTITES, Raf. Fireweed.

E. hieracifolia (L.), Raf.

Wood clearings, especially in recently burned ground; common.

ARCTIUM, L. Burdock.

A. Lappa, L.

Common in waste places, especially about old dwellings. Naturalized from Europe.

CNICUS, L. Thistle.

C. lanceolatus (L.), Hoffm. Common Thistle.

Pastures, fields and roadsides; common. Naturalized from Europe.

C. arvensis (L.), Hoffm. Canada Thistle.

Pastures, fields, and roadsides; too common. Naturalized from Europe.

Forma albiflorus.

Flowers pure white. Southwest Harbor (Rand); — Seal Harbor (Wm. C. Lane).

CENTAUREA, L. Star Thistle.

C. Cyanus, L. Bluebottle. Bachelor's Button.

Occasionally escaped from cultivation to roadsides and waste places, Southwest Harbor, etc. (Rand, Annie S. Downs).

ARNOSERIS, Gærtn. Lamb's Succory.

A small, annual, scapigerous herb; juice milky; leaves all radical; heads few, small; peduncles clavate, fistular; involucral bracts in one series, many, after flowering arching over the fruit ; receptacle flat, naked, pitted ; corollas all ligulate, yellow; anther cells not tailed; upper part of style and its short obtuse arms hairy ; fruit obpyramidal, furrowed and ribbed, not beaked, crowned by a coriaceous angular ring.

A. PUSILLA, Gærtn.

Glabrous or slightly hairy; heads campanulate, 4″ long, inclined in bud; involucral bracts herbaceous, puberulous, linear-lanceolate, tips contracted, obtuse; fruit pale brown, rugose between the ribs; scapes 4′–12′ high, many, slender, rigid, sparingly branched above; leaves 2′–4′ long, narrow, obovate-spatulate or -lanceolate, toothed. Hooker, Fl. Brit. Isles, 229. Rare. Field, Southwest Harbor (M. L. Fernald). Fugitive from Europe.

CICHORIUM, L. CHICORY.

C. INTYBUS, L.

Rare. Formerly in some abundance by roadside, Clark Point, Southwest Harbor (John L. Wakefield, Rand). About 1887 the plant was apparently exterminated in this station. It still persists, however, in another locality on the Point, although in no abundance. Adventive from Europe.

LEONTODON, L. FALL DANDELION.

L. AUTUMNALIS, L.

Fields and roadsides; very common. A form with much aborted ray flowers, Emery District; Southwest Harbor; Great Cranberry Isle (Rand). Naturalized from Europe.

HIERACIUM, L. HAWKWEED.

H. AURANTIACUM, L. FLAMING HAWKWEED.

Fields and meadows; becoming frequent. Beech Hill (R. & R.); — High Head meadow (Rand); — near Ship Harbor (Faxon & Redfield); — near Otter Creek (Theodore G. White). Naturalized from Europe.

H. Canadense, Mx. GREAT HAWKWEED.

Woods and roadsides; frequent.

H. paniculatum, L.

Rare. Clearing on roadside by Denning Pond; east side of Northeast Harbor (Rand).

H. scabrum, Mx. ROUGH HAWKWEED.

Woods and roadsides; common.

PRENANTHES, L. (*Nabalus*, Cass.) RATTLESNAKE-ROOT.

P. serpentaria, Pursh. *Nabalus Fraseri*, DC.

Dry soil; common.

Var. nana (DC.), Gray. *Nabalus nanus*, DC.

Common on mountain summits and rocky places. Green Mt.
(Wm. C. Lane); — Sargent Mt.; Pemetic Mt.; Jordan Mt.;
Beech Cliff; Sutton Island, etc. (Rand).

P. altissima, L. WOOD RATTLESNAKE-ROOT.

Rich damp woods; frequent. A form with dark purple his-
pidulous stems, Southwest Valley road (Rand).

TARAXACUM, Haller. DANDELION.

T. OFFICINALE, Web. *T. Dens-leonis*, Desf. DANDELION.

Becoming common; roadsides and waste places, — sometimes
even in woods. Naturalized from Europe.

LACTUCA, L. LETTUCE.

L. SATIVA, L. GARDEN LETTUCE.

Persistent for years in waste ground, Fernald Point (Rand).
Escaped from cultivation.

L. Canadensis, L. WILD LETTUCE.

Roadside and clearings; frequent.

L. integrifolia, Bigel.

Dry soil; infrequent. Northeast Harbor (John L. Wakefield, Rand); — Somesville; Frenchman Camp (Redfield).

L. leucophæa (Willd.), Gray. BLUE LETTUCE.

Low grounds, roadsides, and waste places; common.

SONCHUS, L. Sow Thistle.

S. oleraceus, L. Common Sow Thistle.

Waste places; rare. Sawyer Cove; Fernald Point (Rand);— High Head (Annie S. Downs);— Clement farm, Seal Harbor (Redfield). Naturalized from Europe.

S. asper, Vill. Spiny Sow Thistle.

Waste places; very common. Naturalized from Europe.

LOBELIACEÆ. Lobelia Family.

LOBELIA, L.

L. cardinalis, L. Cardinal Flower.

Rare and local. Borders of streams, Somesville and vicinity (Rand and others);— also on brook flowing into Seal Cove Pond (Annie S. Downs).

L. spicata, Lam.

Grassy places; frequent. Somesville; Wasgatt Cove (John L. Wakefield);— fields above Long Pond (Redfield);— Seal Harbor (Sara E. Boggs);— Northeast Harbor; Southwest Harbor (Rand);— "Mt. Desert" (F. M. Day).

L. inflata, L. Indian Tobacco.

Dry fields and roadsides; common. Flowers pale blue, violet, or whitish.

L. Dortmanna, L. Water Lobelia.

Common on borders of ponds and often of meadow streams. Usually in shallow water; sometimes immersed.

CAMPANULACEÆ. Campanula Family.

SPECULARIA, Heist. Venus's Looking-glass.

S. perfoliata (L.), A. DC.

Rare. Dry pasture on road to Mason Point, Somesville (R. & R.). Possibly introduced.

CAMPANULA, L. BELLFLOWER.

C. RAPUNCULOIDES, L.

Occasionally by roadsides, etc.; escaped from cultivation. Oak Hill (Redfield, Annie S. Downs); — High Head; Somesville (Rand). Adventive from Europe.

C. rotundifolia, L. BLUEBELL. HAREBELL.

Cliffs on seashore, and frequently on the mountains; common.

Forma **albiflora.**

Flowers white. Ovens (Annie S. Downs); — Otter Cliffs (Annie S. Downs, Rand).

ERICACEÆ. HEATH FAMILY.

GAYLUSSACIA, HBK. HUCKLEBERRY.

G. dumosa (Andr.), T. & G. BOG HUCKLEBERRY.

Frequent in sphagnum bogs. Bog near Somesville (William H. Dunbar); — Somes Pond; Sunken Heath; The Heath, Great Cranberry Isle; Great Heath (Rand).

G. resinosa (Ait.), T. & G. COMMON HUCKLEBERRY.

Dry or wet ground; common.

VACCINIUM, L. BLUEBERRY. CRANBERRY.

V. Pennsylvanicum, Lam. DWARF BLUEBERRY.

Very common everywhere in dry soil; abundant on the hills and mountains. Variable. A form with bluish-red and white fruit, Jordan Mt. (Rand). A well marked form with dark blue-green leaves, reddish shoots, and dark blue fruit with little or no bloom, Great Cranberry Isle (R. & R.).

V. Canadense, Kalm. CANADA BLUEBERRY.

Common in woods or moist ground. Fruit ripening later than that of the preceeding species, and more acid.

V. corymbosum, L. HIGH-BUSH BLUEBERRY.

Swamps and low thickets; frequent. Somesville and vicinity (R. & R.); — Witch Hole (Rand); — Hulls Cove (F. M. Day, J. H. Curtis).

Var. amœnum (Ait.), Gray.

Somesville (Redfield, M. L. Fernald); — Breakneck Ponds (Rand).

This species and variety seem to be rare except in the central and northern parts of the Island.

V. Vitis-Idæa, L. MOUNTAIN CRANBERRY.

Common everywhere, shore and mountains, and on the islands. Fruit much used for sauce, largely taking the place of *V. macrocarpon* for this purpose.

V. Oxycoccus, L. SMALL CRANBERRY.

Common in sphagnum bogs, and in wet places on mountains and shore. Also on Cranberry Isles.

V. macrocarpon, Ait. LARGE CRANBERRY.

Bogs; common, but rarely in great abundance. Also on Cranberry Isles, whence their name.

CHIOGENES, Salisb. CREEPING SNOWBERRY.

C. serpyllifolia, Salisb. *C. hispidula* (L.), T. & G.

Deep mossy woods; common.

ARCTOSTAPHYLOS, Adans. BEARBERRY.

A. Uva-ursi (L.), Spreng. COMMON BEARBERRY.

Open, rocky places; infrequent and local. Browns Mt. (William C. Lane); — Beech Cliff; Dog Mt. (Rand); — Barr Hill; Newport Mt. (Redfield); — "Somes Sound, Southwest Harbor" (Elizabeth G. Britton); — Robinson Mt. (Anna H. Bee); — near Bar Harbor (W. H. Manning).

EPIGÆA, L.   Trailing Arbutus.

E. repens, L.   Mayflower.   Trailing Arbutus.

Frequent in woodlands, but seldom very abundant.

GAULTHERIA, L.   Aromatic Wintergreen.

G. procumbens, L.   Creeping Wintergreen.   Checkerberry.

Common everywhere in woods, shady places, and clearings.

ANDROMEDA, L.

A. polifolia, L.

Bogs; infrequent.   Swampy roadside, south of Salisbury Cove; Hadlock Upper Pond; Sunken Heath (Rand); — Pond Heath (Greenleaf, Lane & Rand); — Great Heath (Redfield).

CASSANDRA, Don.   Leather Leaf.

C. calyculata (L.), Don.

Bogs and marshy borders of ponds; common.

KALMIA, L.   American Laurel.

K. angustifolia, L.   Sheep Laurel.   Lambkill.

Hillsides, pastures, and thickets in dry or damp ground; common.   Also abundant on the mountains.

K. glauca, Ait.   Pale Laurel.

Sphagnum bogs; frequent.   Pond Heath (Greenleaf & Rand, E. Faxon); — Freeman Heath (Faxon & Rand); — Sea Wall Swamp; Sunken Heath; Great Heath (Rand); — bog by roadside west of Sea Wall; The Heath, Great Cranberry Isle (R. & R.).

RHODODENDRON, L.   Rose-bay.   Azalea.

R. Rhodora, Don.   *Rhodora Canadensis*, L.   Rhodora.

Common in damp thickets and swamps, and in wet or even in dry places on the mountains.   Very variable in color of flowers.   Sometimes three to five feet high.

Forma **albiflora.**

Flowers pure white. Southwest Harbor (Annie S. Downs).

LEDUM, L. LABRADOR TEA.

L. latifolium, Ait.

Common in bogs, and often in dry ground. Not very abundant in the southern part of the Island, but common on the Cranberry Isles.

CLETHRA, L. WHITE ALDER.

C. alnifolia, L. SWEET PEPPERBUSH. WHITE ALDER.

Rare. Wet ground near Hadlock Upper Pond (Annie S. Downs). Reported to grow also in meadow on Denning Brook, and on Great Cranberry Isle.

CHIMAPHILA, Pursh. WINTERGREEN.

C. umbellata (L.), Nutt. PRINCE'S PINE. PIPSISSEWA. WINTERGREEN.

Dry woods; frequent.

MONESES, Salisb. ONE-FLOWERED PYROLA.

M. grandiflora, Salisb. *M. uniflora* (L.), Gray.

Deep mossy woods all over the Island; frequent. Also on the Cranberry Isles (Redfield).

PYROLA, L. SHINLEAF.

P. secunda, L. ONE-SIDED PYROLA.

Rich woods; frequent.

P. chlorantha, Swz.

Deep woods; infrequent. Little Harbor Brook Notch; Cold Brook (Rand); — Bar Island, Bar Harbor (F. M. Day); — Seal Harbor (Redfield); — near Beech Hill (Arnold Greene).

P. elliptica, Nutt.

Common in woodlands.

P. rotundifolia, L. Round-leaved Pyrola.

Occasional in dry woods. Great Pond; Salisbury Cove (Henry
C. Jones); — Seal Harbor (Redfield); — Cold Brook (Rand).

MONOTROPA, L. Indian Pipe. Pinesap.

M. uniflora, L. Indian Pipe.

Damp woods; not uncommon.

M. Hypopitys, L. Pinesap.

Dry woods; infrequent. Hadlock Upper Pond; eastern side
of Browns Mt. (William H. Dunbar); — Cold Brook (Rand); —
Green Mt. Gorge (F. M. Day); — Western Mt. (Annie M.
Rand); — Northwest Arm woods (R. & R.).

PLUMBAGINACEÆ. Leadwort Family.

STATICE, L. Marsh Rosemary.

S. Limonium, L., var. **Caroliniana,** Gray. Sea Lavender.

Muddy beaches and salt marshes ; frequent. Somes Harbor,
and shores at head of Somes Sound (Henry C. Jones, R. & R.);
— near Ovens; Bass Harbor; High Head, and northern shores
of the Island (Rand); — Great Cranberry Isle (R. & R.).

PRIMULACEÆ. Primrose Family.

TRIENTALIS, L. Star-flower.

T. Americana (Pers.), Pursh. Star-flower. Star Anemone.

Low woods; common.

LYSIMACHIA, L. Loosestrife.

L. quadrifolia, L. Whorled Loosestrife.

Open woods, hills, and roadsides; common.

L. stricta, Ait. Swamp Loosestrife.

Wet ground and swamps; common.

Var. ovata.

Stems at length much branched above, the branches often equalling or exceeding the inflorescence; leaves ovate to ovate-lanceolate, narrowing less abruptly at the base, shorter-petioled, veins apparent ; raceme short and few-flowered. Wet ground, Somesville (Redfield).

L. thyrsiflora, L. Tufted Loosestrife.

Rare. Intervale Brook, near Hulls Cove (F. M. Day); — bog, northern foot of Beech Hill (Rand).

GLAUX, L. Sea Milkwort.

G. maritima, L.

Salt marshes and muddy beaches; frequent.

OLEACEÆ. Olive Family.

FRAXINUS, L. Ash.

F. Americana, L. White Ash.

Moist woods; common.

F. sambucifolia, Lam. Black Ash.

Swamps and damp woods; frequent.

SYRINGA, L. Lilac.

S. vulgaris, L. Common Lilac.

Occasionally escaped to roadsides near dwellings (R. & R.). Adventive from Eastern Europe or Asia.

APOCYNACEÆ. Dogbane Family.

APOCYNUM, L. Dogbane.

A. androsæmifolium, L. Spreading Dogbane.

Roadsides and thickets; infrequent. Otter Creek (William H. Dunbar); — Northeast Harbor, and elsewhere (Rand); —

Prettymarsh (Redfield); — Somesville ; Hadlock farm, Seal Harbor (R. & R.); — Bar Harbor (F. M. Day).

GENTIANACEÆ. GENTIAN FAMILY.

BARTONIA, Muhl.

B. tenella, Muhl.

Rare. Damp hollows, summit of Green Mt. (Rand, William C. Lane).

MENYANTHES, L. BUCKBEAN.

M. trifoliata, L.

Bogs; rare. Northeast Harbor (John L. Wakefield); — Somes Pond (Rand); — Great Duck Island (Redfield); — Mt. Desert (F. L. Temple).

LIMNANTHEMUM, Gmel. FLOATING HEART.

L. lacunosum (Vent.), Griseb.

Frequent in ponds. Hadlock Lower Pond (William H. Dunbar); — Witch Hole (Rand, F. M. Day, Redfield); — Ripples Pond ; Great Pond ; Denning Pond (Rand); — Eagle Lake ; Mountain Pond; Newport Pond (Redfield).

BORRAGINACEÆ. BORAGE FAMILY.

MERTENSIA, Roth. LUNGWORT.

M. maritima (L.), Don. SEA LUNGWORT.

Frequent on sea beaches, especially on southern coast of the Island, and on the Cranberry Isles. Flowers pink to blue, very rarely white.

LYCOPSIS, L. BUGLOSS.

L. ARVENSIS, L. SMALL BUGLOSS.

Naturalized for years at Fernald Point (William H. Dunbar, Rand); — also in waste ground, Somesville (Rand). Naturalized from Europe.

CONVOLVULACEÆ. Convolvulus Family.

CONVOLVULUS, L. Bindweed.

C. sepium, L., var. Americanus, Sims. Wild Morning Glory.
Sea beaches; common.

CUSCUTA, L. Dodder.

C. Gronovii, Willd.
Infrequent; mostly on the coast. Growing on various plants, especially Aster, Solidago, and Ligusticum. Baker Island (Henry C. Jones, Redfield); — Southwest Harbor; Somesville; Sea Wall, etc. (Rand).

SOLANACEÆ. Nightshade Family.

SOLANUM, L. Nightshade.

S. Dulcamara, L. Bittersweet.
Near dwellings and in low grounds. Hulls Cove; Breakneck Road (F. M. Day); — Bar Harbor (W. H. Manning). Adventive from Europe.

S. nigrum, L. Common Nightshade.
Frequent on sea beaches, seldom elsewhere. Cranberry Isles (John L. Wakefield, R. & R.); — Greening Island; Mill Cove; Bar Harbor; Sea Wall, etc. (Rand). Probably not indigenous within our limits.

NICANDRA, Adans. Apple of Peru.

N. physaloides (L.), Gærtn.
Waste ground, foot of Long Pond (R. & R.). Adventive from South America.

SCROPHULARIACEÆ. Figwort Family.

VERBASCUM, L. Mullein.

V. Thapsus, L. Common Mullein.
Fields, pastures, and roadsides; frequent. Naturalized from Europe.

LINARIA, Juss.   TOAD FLAX.

L. Canadensis (L.), Dumont.   WILD TOAD FLAX.

Frequent in dry soil. An exceedingly depauperate form in gravelly hollows among rocks, especially on the mountains. Browns Mt.; Flying Mt., etc. (Rand); — shore, Northeast Harbor (B. E. J. Gresham); — Baker Island (Redfield).

L. VULGARIS, Mill.   BUTTER-AND-EGGS.

Roadsides; infrequent. Southwest Harbor ; Town Hill; Great Cranberry Isle (Rand); — Baker Island (Redfield).

CHELONE, L.   SNAKE-HEAD.   TURTLE-HEAD.

C. glabra, L.

Wet places, along brooks and rills; frequent.

ILYSANTHES, Raf.

I. riparia, Raf.   *I. gratioloides* (L.), Benth.   FALSE PIMPERNEL.

Rare. Muddy border of Somes Stream (R. & R.); — shore of little mill-pond, Somesville (Rand).

VERONICA, L.   SPEEDWELL.

V. scutellata, L.   MARSH SPEEDWELL.

Boggy ground; infrequent. Northeast Harbor (William H. Dunbar); — Ripples Pond; High Head meadow; bog near Sea Wall (Rand); — "Mt. Desert " (F. M. Day).

V. officinalis, L.   COMMON SPEEDWELL.

Dry ground; rare and local. Roadsides and fields, Salisbury Cove (Faxon, R. & R.); — "Norway Drive," south of Salisbury Cove (Rand, Mary Minot). Apparently confined to the neighborhood of Salisbury Cove, and appearing both introduced and indigenous.

V. serpyllifolia, L.   THYME-LEAVED SPEEDWELL.

Fields, clearings, and roadsides; common. Apparently both introduced and indigenous.

V. peregrina, L. PURSLANE SPEEDWELL.

Dry places; frequent. Flying Mt. (Henry C. Jones); — Norwood Cove, etc. (Rand); — Seal Harbor; Great Cranberry Isle; Great Duck Island (Redfield); — Somesville (M. L. Fernald); — mouth of Denning Brook (R. & R.). All plants small and dwarfed, hardly branched, very unlike the common garden form of this weed.

V. ARVENSIS, L. CORN SPEEDWELL.

Dry places; infrequent. High Head; Flying Mt. (Rand); — Little Cranberry Isle (Redfield); — mouth of Denning Brook (R. & R.); — Bar Harbor (Dr. H. C. Chapman). Appearing indigenous here, but said to be naturalized from Europe.

V. BUXBAUMII, Ten.

Waste ground; rare. Norwood Cove (Rand). Adventive from Europe.

EUPHRASIA, L. EYEBRIGHT.

E. officinalis, L.

Dry ground; common in the southern part of the Island, and on the neighboring islands. Also Bar Harbor (W. H. Manning). If introduced, of very early introduction, but probably indigenous. Very variable. A form from Sea Wall (Rand), having extremely small flowers with corolla scarcely spreading, and leaves less toothed and cut and more crenate, corresponds fairly well to the description of var. *Tatarica*, Benth., but does not agree with herbarium specimens. It appears to be intermediate between *E. curta*, Fries, and *E. gracilis*, Fries. The species, however, is so variable, and has been so subdivided, that it is impossible to name with any certainty its subspecies and varieties without a careful study of the type specimens.

RHINANTHUS, L. YELLOW RATTLE.

R. Crista-galli, L.

Common in fields and on roadsides. If introduced, of very early introduction.

PEDICULARIS, L. Lousewort.

P. Canadensis, L. Common Lousewort.

Fields; common in centre, west, and north of the Island; rare elsewhere, e. g. Beech Hill (R. & R.); — Northeast Harbor; Southwest Harbor (Rand).

MELAMPYRUM, L. Cow Wheat.

M. Americanum, Mx.

Frequent in dry, open woods.

OROBANCHACEÆ. Broom-rape Family.

EPIPHEGUS, Nutt. Beechdrops.

E. Virginiana (L.), Bart.

Under beech trees; rare. Northern end of Jordan Pond (Redfield); — Clark Valley (Rand).

APHYLLON, Mitchell. Naked Broom-rape.

A. uniflorum (L.), Gray. One-flowered Broom-rape.

Low ground, woods and copses; rare. Near Little Harbor; head of The Barcelona meadow (Rand).

LENTIBULARIACEÆ. Bladderwort Family.

UTRICULARIA, L. Bladderwort.

U. inflata, Walt. Floating Bladderwort.

Rare. Witch Hole (Rand, F. M. Day, Redfield).

U. clandestina, Nutt.

Rare. Mountain Pond (Rand); — pools, west side of Great Cranberry Isle (R. & R.).

U. vulgaris, L. Greater Bladderwort.

Marshy ponds, pools, and slow streams; common.

U. gibba, L.

Pond shores; rare. Breakneck Ponds (F. M. Day);—Somes Pond (Rand). A form from mud flats, Somes Pond (Rand), closely approaches *U. biflora*, Lam., and may perhaps prove to be that species. "The flower has the spurs of *U. biflora* very decidedly, but the foliage and the bladders are those of *U. gibba*. The spur here is oblong, narrow, not curved but projecting straightwise, and the perianth is somewhat larger than is generally the case in *U. gibba*. Other specimens with foliage and bladders better represented might show this to be *U. biflora*, but at present it is safer to call it ' *U. gibba* verging towards *U. biflora* in flowers.' " Dr. Thomas Morong *in litt.*

U. intermedia, Hayne.

Bogs and streams; common. Usually sterile; but in flower, Breakneck Ponds (R. & R., E. Faxon).

U. purpurea, Walt. Large Purple Bladderwort.

Ponds; infrequent. Seal Cove Pond (R. & R.);—Aunt Bettys Pond (Rand).

U. resupinata, B. D. Greene. Small Purple Bladderwort.

Pond shores; rare. Breakneck Ponds (F. M. Day);—Ripples Pond (M. L. Fernald).

U. cornuta, Mx. Long-spurred Bladderwort.

Very common on pond shores, in marshes, and sphagnum bogs.

LABIATÆ. Mint Family.

TEUCRIUM, L. Germander.

T. Canadense, L. American Germander. Wood Sage.

Infrequent on banks and in low ground by the shore, at the head of sea beaches. Long Pond (William L. Worcester);— Duck Cove (Rand);— Seal Harbor (Redfield);—Southwest Harbor (Annie S. Downs);—Otter Creek (R. & R.).

MENTHA, L. MINT.

M. VIRIDIS, L. SPEARMINT.

Rare. Runlet and roadside ditch, Seal Harbor (Redfield); — head of Ripples Pond (Rand). Naturalized from Europe.

M. SATIVA, L. WHORLED MINT.

Brooksides; rare. Northeast Harbor; near Carter Nubble (Rand); — Northeast Creek (M. L. Fernald). Naturalized from Europe.

M. ARVENSIS, L. CORN MINT.

Roadside ditches and moist ground; infrequent, but not rare about Southwest Harbor. Also Great Cranberry Isle (R. & R.); — field above Long Pond (Rand). Naturalized from Europe. A tall form with thinner, more sharply serrate leaves, wet roadside by Juniper Cove; brooksides, Norwood Cove (Rand).

M. Canadensis, L. WILD MINT.

Wet places and shady banks; frequent. Also Great Cranberry Isle (Rand); — Little Cranberry Isle (Redfield).

LYCOPUS, L. WATER HOREHOUND.

L. Virginicus, L. BUGLEWEED.

Common in low ground, and often becoming a weed in cultivated grounds. Small, depauperate forms are frequent.

L. sinuatus, Ell. CUT-LEAVED BUGLEWEED.

Common in wet places.

THYMUS, L. THYME.

T. SERPYLLUM, L. CREEPING THYME.

Well established in a field, Bar Harbor (Mary Minot). Adventive from Europe.

SATUREIA, L. SAVORY.

S. HORTENSIS, L. SUMMER SAVORY.

Escaped from cultivation; in field and waste ground, Southwest Harbor (Rand). Adventive from Europe.

HEDEOMA, Pers. AMERICAN PENNYROYAL.

H. pulegioides (L.), Pers. AMERICAN PENNYROYAL.

Common in dry soil.

NEPETA, L. CATNIP.

N. CATARIA, L. CATNIP.

Rare. Roadside near O'Connor Cove ; Thompson Island (Annie S. Downs); — waste ground near dwellings, Fernald Point; Sutton Island (Rand). Adventive from Europe.

N. GLECHOMA, Benth. GROUND IVY. GILL-OVER-THE-GROUND.

Door yards and waste grounds ; infrequent. Somesville; Southwest Harbor (Rand). Adventive from Europe.

SCUTELLARIA, L. SKULLCAP.

S. lateriflora, L. MAD-DOG SKULLCAP.

Wet shady places; infrequent. Marsh, Valley Cove; head of Northeast Creek; High Head meadow (Rand); — Somesville (Annie S. Downs, Rand); — "Mt. Desert" (F. M. Day); — Bar Harbor (W. H. Manning).

S. galericulata, L. COMMON SKULLCAP.

Common on sea beaches and banks by the shore; more rarely on pond shores.

Forma **rosea.**

Flowers rose-color. Beach, Great Cranberry Isle (Rand).

BRUNELLA, L. SELF-HEAL.

B. vulgaris, L. COMMON SELF-HEAL.

Woods, fields, and roadsides; common.

Forma **albiflora.**

Flowers pure white. Meadows above Long Pond (Redfield).

LEONURUS, L.   Motherwort.

L. CARDIACA, L.   Common Motherwort.

Waste places and by dwellings; rare.   Tarr Valley, near Fernald Point (Annie S. Downs); — Oak Hill; Sargent Cove (Rand).   Adventive from Europe.

GALEOPSIS, L.   Hemp Nettle.

G. TETRAHIT, L.   Common Hemp Nettle.

A common weed in waste places and in cultivated grounds. Said to be naturalized from Europe, but appearing indigenous, at least in northern New England.

STACHYS, L.   Hedge Nettle.

S. ARVENSIS, L.   Woundwort.

Rare.   Field, Bar Harbor (Mary Minot).   Adventive from Europe.

PLANTAGINACEÆ.   Plantain Family.

PLANTAGO, L.   Plantain.

P. major, L.   Common Plantain.

Fields and waysides; very common.   Commonly naturalized from Europe.   This plant, however, is found all over the Island in many places so remote from dwellings or cultivated grounds as to lead to the inference that it may be also indigenous. Thick-leaved saline forms are frequent on sea beaches.   Little Harbor; Little Cranberry Isle (Redfield); — Great Cranberry Isle (Rand).

P. LANCEOLATA, L.   Ribgrass.

Waysides and grass lands; formerly rare, but of late years becoming more common.   Mill Cove; Little Harbor; Southwest Harbor (Rand); — Bar Harbor (Rand, W. H. Manning).

P. decipiens, Barn.   Sea Plantain.

Very common in salt marshes, on beaches, and among rocks on the shore.

P. Patagonica, Jacq., var. aristata (Mx.), Gray.

Rare. Field, Southwest Harbor (Annie S. Downs); — Bar Harbor (W. H. Manning). Adventive from the West.

DIVISION III. APETALÆ.

AMARANTACEÆ. AMARANTH FAMILY.

AMARANTUS, L. AMARANTH.

A. RETROFLEXUS, L. AMARANTH PIGWEED.

Roadsides and cultivated grounds; becoming frequent. Bar Harbor ; Southwest Harbor ; Somesville ; Long Pond, etc. (Rand); — Seal Harbor, etc. (Redfield). Adventive from Tropical America.

A. albus, L. TUMBLEWEED.

Cultivated grounds; becoming frequent. Southwest Harbor; Long Pond; Seal Harbor, etc. (Rand); — field on Northeast Creek (M. L. Fernald). Adventive from the South and West.

CHENOPODIACEÆ. GOOSEFOOT FAMILY.

SPINACIA, L. SPINACH.

Diœcious; flowers axillary, glomerate. Staminate flowers in racemose-paniculate clusters, calyx 4–5-parted, the lobes equal. Calyx of pistillate flowers ventricose-tubular, 2–3-toothed; ovary ovoid, styles 4, elongated, filiform, achene included in the turgid indurated calyx, which is often 2–3-horned on the back; seed vertical and compressed; embryo annular, surrounding the' farinaceous albumen. Darlington, Am. Weeds and Useful Plants (Rev. ed.), 274.

S. GLABRA, Mill.

Herbaceous, glabrous throughout; leaves sagittate, sometimes oblong-ovate, entire, acute, slender-petioled ; flowers green ; fruiting calyx solitary, rounded, without prickles, toothed at

the apex. [For a more detailed description, see DC. Prod.,
xiii. 2. 118.] Abundant in waste ground, Somesville (Rand).
Adventive from Asia.

CHENOPODIUM, L. Pigweed.

C. ALBUM, L. Pigweed.

A common weed in cultivated ground, and on sea beaches.
Very variable. Naturalized from Europe.

ATRIPLEX, L. Orache.

A. patulum, L., var. hastatum (L.), Gray.

Sea beaches, salt marshes, etc.; very common and variable.
Both prostrate and erect forms are found.

Var. littorale (L.), Gray.

Rare. Beach, Great Cranberry Isle (Rand).

SALICORNIA, L. Samphire.

S. herbacea, L.

Salt meadows, and muddy shores on the coast; common.

SUÆDA, Forskal. Sea Blite.

S. linearis (Ell.), Moq.

Common on sea beaches.

SALSOLA, L. Saltwort.

S. Kali, L.

Sea beaches; frequent, especially on Cranberry Isles.

POLYGONACEÆ. Buckwheat Family.

RUMEX, L. Dock. Sorrel.

R. Patientia, L. Patience Dock.

Rare. Dry fields near Little Harbor (Redfield). Adventive
from Europe.

R. Britannica, L. Great Water Dock.

Common in swamps and wet places back of sea beaches; less common in wet ground farther inland,— Long Pond meadows; Somesville (Redfield).

R. salicifolius, Weinm. White Dock.

Sea beaches; frequent, especially on Cranberry Isles and western and northern shores of the Island. Also Northeast Harbor; Sea Wall (Rand); — Bar Harbor (W. H. Manning).

R. verticillatus, L. Swamp Dock.

Rare. Marsh on Northeast Creek; Norwood Cove (Rand).

R. crispus, L. Curled Dock.

Common in cultivated and waste ground. Naturalized from Europe.

R. obtusifolius, L. Bitter Dock.

Waste places; rare. Somesville (Rand); — Bar Harbor (W. H. Manning). Adventive from Europe.

R. Acetosella, L. Field Sorrel.

A very common weed in fields and waste places. Naturalized from Europe.

POLYGONUM, L. Knotweed.

P. aviculare, L. Doorweed.

A common weed about dwellings, by roadsides, etc. Variable; erect or prostrate.

P. Raii, Bab.

Stems long, straggling, prostrate; leaves bending towards the stem, elliptic-lanceolate, flat; ochreæ lanceolate, acute, with few distinct simple veins, at length torn; nut smooth, shining, exceeding the perianth. Resembling *P. aviculare* in habit, but *P. maritimum* in fruit. Filaments broader at the base. It varies with smaller leaves and flowers. Babington, Man. Brit. Bot. (4th ed.) 285. Common on sea beaches. This species has been often mistaken for *P. maritimum*, and so reported. So far as known, however, *P. maritimum* is not found within the

limits of this list. The specimen attributed to Mt. Desert in
Bull. Torr. Bot. Club, xix. 362, is now pronounced to be *P.
Raii* (*fide* J. K. Small).

P. lapathifolium, L.
In cultivated grounds; rare. Southwest Harbor (Rand).
Doubtless introduced.

P. Hartwrightii, Gray.
Rare. Bog, Southwest Harbor (Rand).

P. Careyi, Olney.
Rare. Wet ground on wood road, Town Hill (M. L. Fernald).

P. PERSICARIA, L. LADY'S THUMB.
Common in waste and damp places. Variable. Naturalized
from Europe.

P. Hydropiper, L. WATER PEPPER.
Common in wet places.

P. sagittatum, L. ARROW-LEAVED TEAR-THUMB.
Low grounds; common.

P. CONVOLVULUS, L. BLACK BINDWEED.
Waste grounds and beaches; common. Naturalized from
Europe.

P. cilinode, Mx.
Woods and copses, especially in clearings; common.

FAGOPYRUM, Gærtn. BUCKWHEAT.

F. ESCULENTUM, Mœnch.
Waste ground, Southwest Harbor (Rand). Adventive from
Europe.

EUPHORBIACEÆ. SPURGE FAMILY.

EUPHORBIA, L. SPURGE.

E. CYPARISSIAS, L. GRAVEYARD FLOWER.
Escaped from cultivation to roadsides and waste places. Bar
Harbor; Southwest Harbor; High Head (Rand); — Town Hill
(Greenleaf, Lane & Rand); — Somesville (R. & R.). Natural-
ized from Europe.

URTICACEÆ. Nettle Family.

ULMUS, L. Elm.

U. Americana, L. American Elm.

Rare. Fields and roadsides, Eden (Rand). Introduced in other parts of the Island.

URTICA, L. Nettle.

U. gracilis, Ait.

Frequent about and at the heads of sea beaches. Sometimes in waste ground.

* **U. urens, L.**

Rare. Shore of Little Cranberry Isle (Redfield). Adventive from Europe.

PARIETARIA, L. Pellitory.

P. Pennsylvanica, Muhl.

Rare. Shores of Little Cranberry Isle (Redfield). Doubtless naturalized from farther south.

MYRICACEÆ. Sweet Gale Family.

MYRICA, L. Bayberry.

M. Gale, L. Sweet Gale.

Borders of ponds, streams, and wet meadows; common.

M. cerifera, L. Bayberry.

Rocks on coast; common. Also Browns Mt.; Somesville; Mt. Desert Narrows, etc. (Rand).

M. asplenifolia, L. *Comptonia asplenifolia* (L.), Ait. Sweet Fern.

Dry hills, fields, and borders of woods; common.

CUPULIFERÆ. Oak Family.

BETULA, L. Birch.

B. lenta, L. Black Birch. Sweet Birch.
Woods and copses; frequent.

B. lutea, Mx. f. Yellow Birch.
Common in woods.

B. populifolia, Marsh. White Birch. Gray Birch.
Very common in poor soil.

B. papyrifera, Marsh. *B. papyracea,* Ait. Paper Birch. Canoe
Birch.
Common in woods.

ALNUS, L. Alder.

A. viridis (Chaix), DC. Green Alder.
Very common at all altitudes in dry soil.

A. incana (L.), Willd. Speckled Alder.
Common in low grounds.

CORYLUS, L. Hazel-nut.

C. rostrata, Ait. Beaked Hazel-nut.
Frequent in woods and clearings, and by roadsides. North-
west Cove; Emery District; Dog Mt.; Sargent Mt.; Aunt
Mollys Beach ; Somesville, etc. (Rand); — Echo Notch (R.
& R.).

QUERCUS, L. Oak.

Q. rubra, L. Red Oak.
Frequent; widely distributed, but nowhere very abundant;
rare in the south of the Island. Much dwarfed on the mountains. During the early history of the Island oaks were apparently abundant, although probably of this species only. At
a very early day, however, the oak woods began to be felled for
timber. At Somesville, it is said, there was an oak wood on
the shores of Somes Harbor, which attracted the attention of

Abraham Somes, of Gloucester, Mass., when he chanced to sail up the Sound about 1760. He spent the summer in that vicinity making barrel staves, and then returned home with his cargo. The following year he returned to Somes Harbor for the same purpose, and finally, in 1762, built a house on the shore of the harbor near the present steamboat wharf, and began the permanent settlement of Mt. Desert Island.

Q. ilicifolia, Wang. BEAR OAK. BLACK SCRUB OAK.

Rare. Dog Mt. (Rand, Elizabeth G. Britton).

FAGUS, L. BEECH.

F. ferruginea, Ait. AMERICAN BEECH.

Woods; common.

SALICACEÆ. WILLOW FAMILY.

SALIX, L. WILLOW.

S. lucida, Muhl. SHINING WILLOW.

Wet places and borders of brooks and ponds; frequent. Otter Creek Brook; Great Pond; Northwest Cove, etc. (Rand); — Bubble Pond; Jordan Pond; Long Pond meadows, etc. (Redfield).

Forma **latifolia.**

Leaves $1\frac{1}{2}'$ wide, rounded or subacute at base, cuspidate-acuminate. Swamp north of Beech Hill; Southwest Harbor (Rand); — Bubble Pond (R. & R.).

Forma **angustifolia.**

Leaves narrowly lanceolate, tapering to a long point. Long Pond meadows; Thompson Island, etc. (Rand).

S. FRAGILIS, L. CRACK WILLOW. BRITTLE WILLOW.

Bog, Clark Point, Southwest Harbor; Somesville (Rand). Naturalized from Europe.

S. FRAGILIS × ALBA, Wimmer.

Aments leafy-peduncled, slender, loosely flowered; stamens 2, villous at base; scale yellowish, lingulate; capsule very short-pedicelled, conico-cylindrical, glabrous; style very short, stigmas spreading, recurved, 2-lobed; gland embracing the pedicel; leaves broadly lanceolate tapering to a prolonged slender point, glabrous and shining above, silvery-silky beneath when young. Wimmer, Salices Europææ, 133.   Southwest Harbor (R. & R.). All the common large willow trees of the Island probably belong either to *S. fragilis* or to this hybrid.   Both were of early introduction, and have become spontaneous all over the Island. Forms of *S. alba* may also be looked for.

S. rostrata, Richardson.   *S. livida,* Wahl., var. *occidentalis,* Gray.

Very common in either wet or dry soil.

S. discolor, Muhl.   COMMON SWAMP WILLOW.

Very common in low grounds.

S. humilis × discolor, Bebb.

Leaves as broad and large as those of *S. humilis,* but duller green, softly tomentose beneath, and with shorter petioles; the aments thick as those of *S. humilis,* but usually recurved, and the capsules on shorter pedicels.   Bebb, Gray Man., 6th ed., 483.   Swamp, Southwest Harbor (Redfield).

S. humilis, Marsh.   LOW WILLOW.   PUSSY WILLOW.

Common in dry or wet ground everywhere from sea level to mountain summits.   Forms with large leaves may be described as follows: —

Var. **grandifolia,** Anders.   DC. Prod., xvi. 2. 236.

Leaves obovate-oblong 3′–4′ long, 1½′ broad above the middle, shining above.   Appearing in the following forms: —

Forma **obtusifolia.**

Leaves very short-pointed; either smooth and glaucous, or slightly tomentose beneath.   Near Great Pond (R. & R.); — Sutton Island (Rand).

Forma **acuminata.**

Leaves more attenuate-pointed; glaucous and tomentose beneath, generally becoming smooth.   Seal Harbor (Redfield); — wood road to Denning Pond (Rand).

This species and others, especially *S. discolor*, often bear cone-shaped galls on the ends of the branches. These consist of imbricated leaves, and are caused by the deposit of eggs of insects.

S. tristis, Ait. Dwarf Gray Willow.

Rare. Hollows, between Northeast Harbor and Little Harbor (Rand). Perhaps introduced in this locality from beyond our limits.

S. petiolaris, Smith.

Frequent. Somesville; Bass Harbor Marsh; Great Cranberry Isle; Long Pond meadows, etc. (Rand). The type passes into the next variety.

Var. **angustifolia,** Anders. *S. rosmarinifolia* (Herb. Hook.), Barratt & Hooker.

Leaves narrowly lanceolate, almost linear, margin slightly serrulate or subentire, glaucous beneath, at first silky, at length glabrate. DC. Prod., xvi. 2. 234. Common on meadows and heaths. Long Pond meadows; Pond Heath; Northeast Meadow, etc. (R. & R.). The silky hairs of the young leaves are usually of a rusty color.

S. cordata, Muhl. Heart-leaved Willow.

Rare. Roadside near Denning Pond; Southwest Harbor (Rand).

S. balsamifera (Hook.), Barratt. Balsam Willow.

Widely distributed, but nowhere very abundant. Southwest Harbor; Sargent Mt.; Beech Hill; Ripples Pond; Somes Pond; Denning Brook; Beech Mt. Notch; The Hio; Doctors Brook; road, west side of Browns Mt., etc. (Rand); — Seal Harbor (Redfield); — Browns Mt. (E. Faxon); — High Head meadow (Faxon & Rand); — Long Pond meadows (R. & R.).

POPULUS, L. Poplar. Aspen.

P. tremuloides, Mx. Aspen.

Frequent in woods.

P. grandidentata, Mx. Large-toothed Aspen.

Frequent in woods.

P. balsamifera, L. Balm of Gilead.

Frequent about dwellings, etc. Apparently introduced, and naturalized by seedlings. Oak Hill, etc. (Rand); — Southwest Harbor (Elizabeth G. Britton, Rand); — Seal Harbor (Redfield); — and elsewhere. At Somesville, however, and east of Town Hill, remote from dwellings, it appears to be indigenous.

P. DILATATA, L. Lombardy Poplar.

Occasionally by roadsides, etc., and sometimes spontaneous, Somesville; Southwest Harbor (Rand); — Oak Hill (Redfield). Adventive from Europe.

EMPETRACEÆ. Crowberry Family.

EMPETRUM, L. Crowberry.

E. nigrum, L. Black Crowberry.

Common on cliffs along the coast and on the islands. Often in heaths and bogs; The Heath, Great Cranberry Isle; Sunken Heath; Great Heath (Rand); — more rarely on mountains and hills; Barr Hill; Green Mt. (Redfield); — Sargent Mt. (Rand). Also Duck Islands (Redfield).

COREMA, Don. Broom Crowberry.

C. Conradii, Torr.

Dry rocky places; rare and local. Asticou; Little Harbor (Lane & Rand); — Barr Hill, in a number of localities; hill between Long Pond and Little Harbor Brook (Redfield); — Beech Mt.; Dog Mt. (Rand); — Beech Hill (Annie S. Downs, Sara E. Boggs); — Ship Harbor (Annie S. Downs); — in great abundance in pine barrens, east of Ship Harbor; west of Hio (Rand).*

---

* Corema also grows in abundance with *Pinus Banksiana*, near Prospect Harbor, Gouldsborough (Redfield).

CLASS II. DICOTYLEDONES GYMNOSPERMEÆ.

CONIFERÆ. PINE FAMILY.

PINUS, L. PINE.*

P. Strobus, L. WHITE PINE.
Frequent.

P. rigida, Mill. PITCH PINE.
Barren soil; not uncommon, but local. Browns Mt., etc.
(R. & R.). Very abundant on Newport Mt.

P. resinosa, Ait. RED PINE. NORWAY PINE.
Frequent.

PICEA, Link. SPRUCE.

P. nigra (Ait.), Link. *Abies nigra*, Poir. BLACK SPRUCE.
Common.

P. alba (Ait.), Link. *Abies alba*, Mx. WHITE SPRUCE.
Common. More common near the coast than the preceding species. A very beautiful glaucous or blue form is not uncommon in different parts of the Island.

TSUGA, Carr. HEMLOCK.

T. Canadensis (L.), Carr.
Infrequent, except in old woods. Also on Cranberry Isles.

ABIES, Juss. FIR.

A. balsamea (L.), Mill. BALSAM FIR.
Common.

LARIX, Adans. LARCH.

L. Americana, Mx. HACKMATACK. TAMARACK.
Common. Also on Cranberry Isles.

---

* *P. Banksiana*, Lamb., is found in abundance on Schoodic Peninsula, across Frenchman Bay, but has not yet been found on Mount Desert Island.

THUJA, L.   ARBOR-VITÆ.
T. occidentalis, L.
Very common.   Usually called White Cedar on the Island.

JUNIPERUS, L.   JUNIPER.
J. communis, L.   COMMON JUNIPER.
Common in dry ground, on rocky cliffs, etc.   Very low-spreading, and variable.

J. Sabina, L., var. procumbens, Pursh.   CREEPING JUNIPER.
Frequent in dry fields or on rocky cliffs along the coast.
Roberts Point, Northeast Harbor; Pierce Head; Sargent Mt.
(Rand);—shores Seal Harbor (Redfield);—Bar Harbor (Beatrix Jones);—Bald Porcupine Island (W. H. Manning);—
very common in dry fields, Sutton Island (Rand et als.).

TAXUS, L.   YEW.
T. Canadensis, Willd.   YEW.   GROUND HEMLOCK.
Common in deep moist woods and glens.

CLASS III.   MONOCOTYLEDONES.

ORCHIDACEÆ.   ORCHID FAMILY.

MICROSTYLIS, Nutt.   ADDER'S MOUTH.
M. ophioglossoides (Willd.), Nutt.
Not infrequent, and very generally distributed over the Island
and the Cranberry Isles, in low moist ground.   Also Sargent
Mt.; summit of Flying Mt. (Rand).

LIPARIS, Richard.   TWAYBLADE.
L. Lœselii (L.), Richard.
Local.   On Stanley Brook, Seal Harbor (G. Hunt);—swamp
on Breakneck road (Brigham);—near Kings Point, and on
Clark Point, Southwest Harbor; Somesville; field, Juniper
Cove (Rand).

CORALLORHIZA, R. Br. CORAL-ROOT.

C. innata, R. Br.

Deep woods; infrequent. Hadlock Upper Pond (H. C. Jones, Rand); — Southwest Valley; Cold Brook; Little Harbor Brook Notch ; Clark Valley (Rand); — Sargent Mt. (Brigham); — woods, Jordan Pond road (Redfield); — woods, off Town Hill road, Somesville (R. B. Worthington).

C. multiflora, Nutt.

Infrequent, and not as widely distributed as the last. Hadlock Valley (Rand, Redfield); — Bar Island, Bar Harbor (F. M. Day); — woods, Hadlock Upper Pond; Cold Brook; Clark Valley (Rand).

LISTERA, R. Br. TWAYBLADE.

L. cordata (L.), R. Br.

Rare. Cold Brook (Rand).

L. convallarioides (Swz.), Nutt.

Very rare. Woods, head of The Barcelona meadow (Rand).

SPIRANTHES, Richard. LADIES' TRESSES.

S. Romanzoffiana, Cham.

Frequent in damp ground or meadows. Flowering in late July and August, earlier than the next.

S. cernua (L.), Richard.

Very common in damp ground. Flowering in late August, September, and early October.

S. gracilis (Bigel.), Beck.

Infrequent. Roadside near Little Harbor (W. C. Lane); — Beech Cliff ; roadsides, Northeast Harbor, and near Fernald Cove; Norwood Road, Southwest Harbor (Rand); — Barr Hill; fields north of Long Pond (Redfield); — Southwest Harbor (Annie S. Downs).

GOODYERA, R. Br.   RATTLESNAKE PLANTAIN.

G. repens (L.), R. Br.

Frequent in deep woods.   Also on Cranberry Isles.

G. pubescens (Willd.), R. Br.

Rare.   Sargent Mt. Gorge ; Cold Brook (Rand).

ARETHUSA, L.

A. bulbosa, L.

Common in sphagnum bogs.   A form with two scapes, each two-flowered, Sea Wall Swamp; meadow on Denning Brook (Rand).

Forma **albiflora.**

Flowers pure white.   Sea Wall Swamp (Redfield, Faxon & Rand); — Somes Pond Swamp; Sunken Heath (Rand); — Great Heath, etc. (Annie S. Downs).

Forma **subcærulea.**

Flowers bluish or lavender in color.   Swamp on Denning Brook (Rand).

CALOPOGON, R. Br.

C. pulchellus (Willd.), R. Br.

Common, usually in sphagnum bogs.

POGONIA, Juss.

P. ophioglossoides (L.), Ker.

Common in sphagnum bogs and on wet pond shores.   On the Island this and the two plants last named are called indiscriminately by the name of Swamp Pink.

Forma **albiflora.**

Flowers pure white.   Jordan Pond (Rand); — Sea Wall; Great Heath (Annie S. Downs).

HABENARIA, Willd.

H. tridentata (Willd.), Hook.

Common in moist ground.

H. hyperborea (L.), R. Br.

Rare. Swamp, Northeast Harbor (Rand); — trail between Jordan Pond and Northeast Harbor (Redfield).

H. dilatata (Pursh), Gray.*

Frequent in swampy ground.

H. obtusata (Pursh), Richardson.

Deep, mossy woods; infrequent. Sargent Mt. Gorge; woods, head of The Barcelona meadow; Cold Brook (Rand); — Seal Harbor (Redfield); — Little Cranberry Isle (Redfield, Harriet A. Hill).

H. Hookeri, Torr. †

Rare. Woods near Otter Creek (Helen B. Walley).

H. orbiculata (Pursh), Torr.

Deep woods; infrequent. Sargent Mt. Gorge; Northwest Arm woods (Rand); — woods, southern end of Great Pond (Annie S. Downs); — Seal Harbor (Redfield, Lizzie Churchill); — Little Cranberry Isle (Redfield, Harriet A. Hill).

H. fimbriata (Ait.), R. Br. PURPLE-FRINGED ORCHIS.

Common in swampy ground. A small-flowered form of this species has sometimes been taken for *H. psycodes,* a species not as yet discovered on the Island.

Forma **albiflora.**

Flowers pure white. Hadlock Upper Pond (B. E. J. Gresham); — Long Pond meadows (Redfield).

CYPRIPEDIUM, L. LADY'S SLIPPER.

C. spectabile, Salisb. SHOWY LADY'S SLIPPER.

Swamp, Northeast Harbor (Rand).

---

* Not of Hooker. See A. Gray, Ann. Lyc. N. Y., iii. 231.
† H. Hookeriana, Torr., is the more correct name.

C. acaule, Ait. STEMLESS LADY'S SLIPPER.
Common in woods.
Forma **albiflorum.**
Flowers pure white, or white with indistinct pink veins.
Near Breakneck Ponds (Rand); — Beech Mt. Notch; woods,
Hadlock Lower Pond (R. & R.).

IRIDACEÆ. IRIS FAMILY.

IRIS, L. BLUE FLAG.

I. versicolor, L.
Very common everywhere in moist ground, especially in low
grounds near the coast.

SISYRINCHIUM, L. BLUE-EYED GRASS.

S. angustifolium, Mill.   *S. Bermudiana,* L.*
Common in grassy places. Whatever may be the fact in regard to the specific rank of *S. anceps* and *S. mucronatum,* all the
Island forms must be referred to *S. angustifolium.* A specimen
with a single spathe, collected by F. M. Day, in 1882, probably
in the vicinity of Bar Harbor, has been somewhat doubtfully
marked *S. anceps* in the herbarium. A recent and more careful
examination, however, seems to show that it is nothing more than
S. angustifolium with smaller, probably immature seeds. It is
worthy of remark that not a specimen of the true *S. anceps* form
has yet been found on the Island, although *S. angustifolium* is
so very abundant. This fact is evidence that these forms of the
plant merit at least varietal distinction.

LILIACEÆ. LILY FAMILY.

HEMEROCALLIS, L. DAY LILY.

H. FULVA, L.
Often by roadsides near dwellings; escaped from cultivation.
Town Hill (R. & R.); — Somesville; Emery District; South-
west Harbor (Rand). Adventive from Europe.

* See Morong, Bull. Torr. Bot. Club, xx. 467.

POLYGONATUM, Adans. SOLOMON'S SEAL.

P. biflorum (Walt.), Ell. SMALLER SOLOMON'S SEAL.

Infrequent. Head of Little Harbor Brook Notch; woods, Deer Brook ; Canada Valley; copses, Long Pond meadows (Rand); — Hadlock Valley (Redfield); — Rum Key, Porcupine Islands (F. M. Day).

ASPARAGUS, L.

A. OFFICINALIS, L. GARDEN ASPARAGUS.

Escaped from cultivation; in uncultivated field, Southwest Harbor (Rand). Adventive from Europe.

SMILACINA, Desf. FALSE SOLOMON'S SEAL.

S. racemosa (L.), Desf. FALSE SPIKENARD.

Infrequent. Head of Northeast Harbor; woods, Deer Brook; Northwest Arm woods; Southwest Valley road (Rand); — Hadlock Valley; Browns Mt. Notch (Redfield); — Rum Key, Porcupine Islands (F. M. Day).

S. stellata (L.), Desf.

Very rare. Porcupine Islands (F. M. Day).

S. trifolia (L.), Desf.

Frequent in peat bogs.

MAIANTHEMUM, Wiggers.

M. Canadense, Desf. *Smilacina bifolia*, Ker., var. *Canadensis*, Gray. DWARF SOLOMON'S SEAL.

Common everywhere in woods, copses, and clearings.

STREPTOPUS, Mx. TWISTED STALK.

S. amplexifolius (L.), DC.

Frequent in deep woods, especially along mountain brooks.

S. roseus, Mx.

Frequent in deep woods. Perhaps more common than the last.

CLINTONIA, Raf.   WILD LILY OF THE VALLEY.

C. borealis (Ait.), Raf.

Common in deep, moist woods.

OAKESIA, S. Wats.

O. sessilifolia (L.), S. Wats.   *Uvularia sessilifolia*, L.   COMMON
BELLWORT.

Common in deciduous woods and in copses.

ERYTHRONIUM, L.   DOG-TOOTH VIOLET.

E. Americanum, Ker.

Rare.   Low ground, Clark Point, Southwest Harbor; Fernald
Point (Anna H. Bee).

LILIUM, L.   LILY.

L. Philadelphicum, L.   BLACKBERRY LILY.   WILD RED LILY.

Local, and not widely distributed.   Schooner Head (Rand) ; —
Sargent Mt. ;   Somesville (R. & R.); — Norway Drive, and
country about Salisbury Cove (F. M. Day, Clara L. Walley,
Mary Minot, Margaret A. Rand).

L. Canadense, L.   WILD YELLOW LILY.   CANADA LILY.

Rare.   Salisbury Cove (Mary Minot).   Very common, however, on the mainland.

MEDEOLA, L.   INDIAN CUCUMBER ROOT.

M. Virginiana, L.

Frequent in rich woods.

TRILLIUM, L.   WAKE ROBIN.

T. erythrocarpum, Mx.   PAINTED TRILLIUM.

Frequent in damp woods.

PONTEDERIACEÆ. Pickerel-weed Family.

PONTEDERIA, L. Pickerel-weed.

P. cordata, L.

Streams and muddy pond shores; common. Forms are sometimes found corresponding to var. *angustifolia* (Pursh), Torr.

XYRIDACEÆ. Yellow-eyed Grass Family.

XYRIS, L. Yellow-eyed Grass.

X. flexuosa, Muhl., var. pusilla, Gray.

Peat bogs and sandy shores; rare and local. Hadlock Ponds (Wm. H. Dunbar, Rand); — Breakneck Ponds (A. H. Smith, R. H. Day); — Jordan Pond (T. G. White, Redfield, Annie M. Rand). It is possible that this may be distinct from *X. flexuosa*, and entitled to specific rank. See Bull. Torr. Bot. Club, xix. 38, where this plant is described as *Xyria montana*, Ries, — *Xyria* being an evident typographical error for *Xyris*. The specific name *montana*, there chosen, is most unfortunately inappropriate.

JUNCACEÆ. Rush Family.

JUNCUS, L. Rush.

J. effusus, L. Common Rush.

Very common in wet ground.

J. filiformis, L.

Rare. Edge of pool on Heaths Brook, Tremont, near head of the Marsh (M. L. Fernald).

J. Balticus, Dethard, var. littoralis, Engelm.

Common in wet brackish ground on the coast. Also on White Beach, southern end of Great Pond (Rand).

J. Greenei, Oakes & Tuck.

Dry ground; rare. Sargent Mt. (Rand); — Southwest Harbor (M. L. Fernald).

J. tenuis, Willd.

Fields and roadsides; very common.

Var. **secundus** (Beauv.), Engelm.

Rare. Dog Mt. (Rand).

J. Gerardi, Loisel. BLACK GRASS.

Salt marshes; common.

J. bufonius, L.

Low ground, especially by roadsides; very common.

J. pelocarpus, E. Meyer.

Wet meadows and pond shores; common. A proliferous form, Denning Pond (Rand).

J. articulatus, L.

Wet grounds and roadsides; frequent. Northeast Harbor (Greenleaf, Rand); — Somesville (Redfield); — Southwest Harbor; High Head (Rand).

J. militaris, Bigel.

Common in streams and ponds.

J. acuminatus, Mx.

Mt. Kebo (Greenleaf); — near Long Pond (Redfield). Probably not uncommon, but as yet seldom reported.

J. Canadensis, J. Gay, var. longicaudatus, Engelm.

Frequent. Great Head (Redfield); — Southwest Harbor; Hadlock Upper Pond; Denning Pond (Rand); — Cranberry Isles; Somesville (R. & R.).

Var. **coarctatus,** Engelm.

Very common.

LUZULA, DC. WOOD RUSH.

L. vernalis, DC. *L. pilosa* (L.), Willd.

Infrequent. Clearing north of Pond Heath; meadow between Somesville and Town Hill (Rand); — clearings on Indian Point road, Somesville (R. & R.).

L. campestris (L.), DC.

Common in fields, clearings, etc.

TYPHACEÆ. Cat-tail Family.

TYPHA, L. Cat-tail Flag.

T. latifolia, L.

Bogs and marshes; frequent.

SPARGANIUM, L. Bur-reed.

S. simplex, Huds.

Common in shallow water, in brooks and ditches. Very variable.

Var. **androcladum, Engelm.**

Common. From a study of Mt. Desert specimens it appears better not to recognize this variety as entitled to specific rank. There is no well defined dividing line between it and *S. simplex,* and occasionally the two forms appear on the same plant.

Var. **fluitans, Engelm.**

Floating in water of moderate depth. Somes Pond; Southeast Creek, Bass Harbor (Rand); — Witch Hole (Redfield).

Var. **angustifolium (Mx.), Engelm.**

Rare. Seal Cove Pond (R. & R.); — pool, Hunters Brook (Rand). Apparently also in Hadlock Upper Pond (Rand).

S. minimum, Fries.

Ponds and streams; frequent. The Barcelona; Hadlock Upper Pond, etc. (Rand); — Jordan Stream (R. & R.); — Northeast Creek (M. L. Fernald).

ARACEÆ. Arum Family.

ARISÆMA, Mart. Indian Turnip.

A. triphyllum (L.), Torr. Jack-in-the-Pulpit.

Rich woods, boggy places, and meadows; infrequent. Jordan Pond trail from Northeast Harbor; The Barcelona meadow; Northeast Meadow (Rand); — roadside, north of Doctors Creek (R. & R.).

CALLA, L.   WATER ARUM.

C. palustris, L.   WILD CALLA LILY.

Rare.   Near Northeast Harbor (F. L. Temple); — The Heath, Great Cranberry Isle (Arnold Greene, Redfield).

SYMPLOCARPUS, Salisb.   SKUNK CABBAGE.

S. fœtidus (L.), NUTT.

Swamps ; rare.   Cranberry Isles (H. C. Jones, Rand, Redfield); Sea Wall and vicinity (Rand).   So far as at present known the distribution of this species is very peculiar.   There seems to be no good reason why it should be confined to such a limited area.   It may be noted that a slight elevation of the coast line would connect the Cranberry Isles, where it occurs most frequently, with Mt. Desert Island at the Sea Wall and vicinity, where it also occurs.

ACORUS, L.   SWEET FLAG.

A. Calamus, L.

Low grounds ; rare.   Near head of Doctors Creek ; near Otter Creek (Redfield); on the "Overflow Brook," Somesville (R. & R.).

ALISMACEÆ.   WATER-PLANTAIN FAMILY.

SAGITTARIA, L.   ARROWHEAD.

S. variabilis, Engelm.

Common in shallow water or wet places.   Very variable; the various so-called varieties or forms apparently passing into one another by intermediate forms.

Forma **hastata.**

Common.

Forma **obtusa.**

Somes Stream (W. H. Dunbar, R. & R.).

Forma **angustifolia.**

Near Bar Harbor (F. M. Day); — Seal Cove Pond; Ripples Pond (Rand); — Somesville (R. & R.).

S. graminea, Mx.

Infrequent. Somes Stream (W. H. Dunbar, R. & R., Arnold Greene); — Great Pond; Seal Cove Pond (Rand).

NAIADACEÆ. Pondweed Family.

TRIGLOCHIN, L. Arrow Grass.

T. maritima, L.

Common in salt marshes, and on muddy coast shores.

SCHEUCHZERIA, L.

S. palustris, L.

Rare. Sunken Heath (Faxon, Rand); — The Heath, Great Cranberry Isle (R. & R.).

POTAMOGETON, L. Pondweed.

P. natans, L.

Ponds and deep streams; frequent. Seal Cove Pond; Great Pond (Rand); — Somes Pond (R. & R.); — Northeast Creek; Witch Hole (Redfield).

P. Oakesianus, Robbins.

Rare. Northeast Creek; Bubble Pond (R. & R.).

P. Pennsylvanicus, Cham. *P. Claytonii,* Tuck.

Common in ponds, streams, and ditches. A very slender deep-water form, Somes Pond (Rand).

P. hybridus, Mx. *P. diversifolius,* Raf.

Rare. Ripples Pond (Fernald, Rand). With and without floating leaves.

P. perfoliatus, L.

Infrequent. Long Pond; Jordan Stream (Redfield); — Northeast Creek (E. Faxon, Rand, M. L. Fernald).

11

Var. lanceolatus, Robbins.   Var. *Richardsonii*, Bennett.
Rare.   Northwest Arm, Great Pond (M. L. Fernald).
P. pusillus, L.
Rare.   Northeast Creek (M. L. Fernald).

RUPPIA, L.   DITCH GRASS.

R. maritima, L.
Ponds and streams of brackish water along coast; common.

ZOSTERA, L.   EEL GRASS.

Z. marina, L.
Very common in shoal water along the coast.

NAIAS, L.   NAIAD.

N. flexilis (Willd.), Rostk. & Schmidt.
Rare.   Ripples Pond (Rand).

ERIOCAULEÆ.   PIPEWORT FAMILY.

ERIOCAULON, L.   PIPEWORT.

E. septangulare, With.
Common in ponds and along pond borders.   Sometimes in deep water, sending to the surface scapes six to ten feet in length, as in Jordan Pond (R. & R.).

CYPERACEÆ.   SEDGE FAMILY.

DULICHIUM, Pers.

D. spathaceum (L.), Pers.
Frequent on borders of ponds and streams.

ELEOCHARIS, R. Br.   SPIKE RUSH.

E. ovata (Roth), R. Br.   *E. obtusa*, (Willd.), Schultes.
Uncommon.   Muddy margins of mill-pond and of Somes Stream, Somesville; bog-hole, High Head; stream near head of Southwest Harbor (Rand).

E. olivacea, Torr.

Rare. Sandy margin of Somes Stream (M. L. Fernald).

E. palustris (L.), R. Br.

Swamps; infrequent. Seal Harbor; Baker Island (Redfield).

Var. **glaucescens** (Willd.), Gray.

Very common in marshy ground along the coast. Southwest Harbor; Sea Wall; Somesville; Great Cranberry Isle, etc. (Rand); — Great Duck Island (Redfield).

E. tenuis (Willd.), Schultes.

Wet ground and pond shores; very common. Variable.

E. acicularis (L.), R. Br.

Muddy places; infrequent. Hadlock Lower Pond (Greenleaf); — Somesville (R. & R.).

E. pygmæa, Torr.

Rare. Brackish marsh, Little Cranberry Isle (Redfield).

SCIRPUS, L. CLUB RUSH.

S. cæspitosus, L.

Local and infrequent. Boggy depressions, Sargent Mt. (R. & R.); — The Heath, Great Cranberry Isle, — in great abundance; Sunken Heath (Rand).

S. subterminalis, Torr.

Ponds and streams; infrequent. Bar Harbor (W. Boott); — Seal Cove Pond and Brook; Somes Pond (R. & R.); — Denning Pond and Brook; meadow at outlet of Great Pond (Rand).

S. pungens, Vahl.

Rare. Bog on shore south of Sea Wall (Rand).

S. lacustris, L. *S. validus,* Vahl.

Ponds and marshes; common. In brackish water, Sea Wall (Rand); — Great Cranberry Isle (R. & R.); — Little Cranberry Isle (Redfield).

S. maritimus, L.

Salt marshes and muddy beaches along the coast ; very common.

S. sylvaticus, L., var. **digynus,** Boeckl.   *S. microcarpus,* Presl.

Wet ground; common.   Blooming earlier than the next.

S. atrovirens, Muhl.

Wet ground ; frequent.   Northeast Harbor;   High Head; Prettymarsh ;   Southwest Harbor;   Seal Cove, etc. (Rand); — Seal Harbor (Redfield).

ERIOPHORUM, L.   COTTON GRASS.

E. cyperinum, L.   *Scirpus Eriophorum,* Mx.

Wet ground; common and variable.

Var. **laxum** (Gray), Wats. & Coult.

Frequent.   Seal Harbor; Long Pond meadows; Bubble Pond (Redfield); — Southwest Harbor (Rand).

E. alpinum, L.

Rare.   Bog, Northeast Harbor (R. & R.); — borders of Upper Breakneck Pond (Redfield).

E. vaginatum, L.

Peat bogs and swamps ; frequent.   Southwest Harbor; Sargent Mt. (Greenleaf, Lane & Rand); — Little Cranberry Isle ; Great Heath (Redfield) ; — Sunken Heath, etc. (Rand).

E. Virginicum, L.

Open bogs and swamps; common.   A form bearing unequally peduncled spikelets, Great Cranberry Isle (Rand).

E. polystachyon, L.

Frequent.   Meadow on Sunken Heath Brook (Redfield); — Sea Wall Swamp (Rand); — Prettymarsh (Greenleaf, Lane & Rand).

Var. **latifolium** (Hoppe), Gray.

More common than the type, — if the varietal distinction is valid.   Sargent Mt. (Greenleaf, Lane & Rand); — Hulls Cove;

Somesville, etc. (Rand); — Seal Harbor, etc.; Cranberry Isles (Redfield).

E. gracile, Koch.

Frequent. Seal Harbor; Little Cranberry Isle (Redfield); — Southwest Harbor; High Head; Great Cranberry Isle (Rand); — "Mt. Desert" (F. M. Day).

RHYNCHOSPORA, Vahl. BEAK RUSH.

R. fusca (L.), Roem. & Schultes.

Frequent. Long Pond meadows; Sea Wall (Redfield); — Somes Pond Swamp (R. & R., Fernald); — Denning Brook; at outlet of Great Pond; Ripples Pond, etc. (Rand).

R. alba (L.), Vahl.

Wet ground; common.

CLADIUM, P. Br. TWIG RUSH.

C. mariscoides (Muhl.), Torr.

Infrequent. Outlet of Hadlock Upper Pond (R. & R.); — "Mt. Desert" (F. M. Day); — Somesville (M. L. Fernald); — swamp above Long Pond (Redfield).

CAREX, L. SEDGE.

C. pauciflora, Lightf.

Rare. Borders of Sea Wall Swamp (R. & R.); — Sunken Heath (Rand).

C. Michauxiana, Boeckl. *C. rostrata,* Mx. *C. abacta,* Bailey.

Rare. Bog, Somes Pond (Rand).

C. folliculata, L.

Common in swamps and damp meadows.

C. intumescens, Rudge.

Common in swamps and wet ground.

C. oligosperma, Mx.

Rare. "Meadow at Gorge, near Bar Harbor" (W. Boott, spec. in Gray Herb.); — "Mt. Desert" (Randall Spaulding); — marsh, foot of Lower Breakneck Pond (E. Faxon).

C. utriculata, Boott.

Swamps; common. A form with very narrow spikes, Pond Heath (R. & R.).

Var. minor, Boott.

Infrequent. Long Pond meadows (Redfield); — Northeast Meadow (R. & R.).

C. lurida, Wahl.   *C. tentaculata*, Muhl.

Common in swamps and wet ground everywhere.

C. hystricina, Muhl.

Rare. Meadow on Little Harbor Brook (Redfield).

C. Pseudo-Cyperus, L.

Not uncommon. Seal Harbor (Redfield); — bog south of Sea Wall; bog at northern foot of Beech Hill (Rand).

C. scabrata, Schw.

Common in wet ground.

C. Houghtonii, Torr.

Frequent by roadsides and in clearings. Roadside, Intervale Brook valley; Sea Wall (R. & R.); — clearings, Youngs District (Faxon & Rand); — East Peak of Western Mt.; near Oak Hill; road west of Browns Mt. (Rand); — Southwest Harbor (Faxon); — Bubble Pond road (T. G. White).

C. filiformis, L.

Common in bogs and on boggy shores of ponds and streams. An immature form, probably var. *latifolia*, by roadside south of Salisbury Cove (Greenleaf, Lane & Rand).

C. fusca, All.   *C. Buxbaumii*, Wahl.

Rare. Summit of Green Mt. (E. Faxon).

C. vulgaris, Fries. *C. rigida,* Good., var. *Goodenovii* (J. Gay), Bailey (Gray, Man., 6th ed., 2d issue, 735 *c*).

Common in low grounds; variable. A singular dwarf form, with small very black spikes, Sea Wall (Rand); — Somesville (Faxon & Rand).

Var. **strictiformis,** Bailey. *C. rigida,* Good., var. *strictiformis,* Bailey, *l. c.*

Frequent. Northeast Harbor (Greenleaf); — Somesville (R. & R.); — Great Cranberry Isle (Rand); — Northwest Arm, Great Pond; Little Cranberry Isle (Redfield).

C. stricta, Lam.

Common in wet ground.

Var. **angustata** (Boott), Bailey.

Infrequent. Intervale Brook valley (Rand); — Somesville and vicinity (Redfield, Faxon, Rand).

Var. **decora,** Bailey.

Rare. Meadow on Sunken Heath Brook (Rand).

C. lenticularis, Mx.

Gravelly borders of ponds and streams; frequent. Stanley Brook, Seal Harbor (Redfield); — Jordan Pond; Great Pond (R. & R.).

C. maritima, O. F. Mueller.

Marshy shores on the coast; common. Southwest Harbor; Northeast Harbor; Seal Harbor; Otter Creek; Thomas Bay; High Head; Seal Cove; Bass Harbor, etc.

C. crinita, Lam.

Low ground; common.

Var. **minor,** Boott.

Rare. On Little Harbor Brook, at crossing of Northeast Harbor trail to Jordan Pond (Redfield).

C. Magellanica, Lam. *C. irrigua,* Smith.

Frequent in cold bogs. Also Green Mt. (Redfield); — Sargent Mt. (R. & R.); — Little Cranberry Isle (Redfield).

C. arctata, Boott.

Common in woodlands.

C. debilis, Mx., var. Rudgei, Bailey.

Common.

C. gracillima, Schw.

Meadows and low ground; frequent. Seal Harbor (Redfield);
— south of Town Hill; High Head meadow (Rand); — Northeast Meadow (R. & R.), and elsewhere.

C. flava, L.

Common in low ground.

Var. **graminis, Bailey.**

Uncommon. Shores, Northwest Arm, Great Pond; Ripples
Pond (Rand); — field near Ship Harbor (Redfield & Faxon); —
Long Pond meadows (Redfield). Depauperate forms of this va-
riety are found in abundance on shores of Great Pond (Rand).

Var. **viridula** (Mx.), Bailey. *C. Œderi*, Gray, Man., 5th ed.

Common in both wet and dry ground.

C. pallescens, L.

Common in low ground.

C. conoidea, Schk.

Low ground about Long Brook, Great Cranberry Isle (Redfield, Rand); — Southwest Harbor (M. L. Fernald).

C. laxiflora, Lam.

Grassy places; common and variable.

Var. **varians, Bailey.**

With the type, and as common.

C. PANICEA, L.

Rare. Low grassy ground at mouth of Long Brook, Great
Cranberry Isle (Theodore G. White, Redfield). Appearing
indigenous, but probably naturalized here from Europe for many
years.

C. deflexa, Hornem. *C. Novæ-Angliæ,* Gray, Man., 5th ed., mostly; *non* Schweinitz.

Not uncommon on mountains, and in dry ground in woods and clearings. Sargent Mt.; Beech Mt. Notch; Breakneck road; Southwest Valley road; Southwest Harbor; Hadlock Lower Pond; Somesville (Rand). On Town Hill road, Somesville, a form approaching var. *Deanei* (Rand).

Var. Deanei, Bailey.

Frequent in dry clearings. Somes Pond pastures ; Hulls Cove; Beech Hill cross-road (Rand); — Somesville; Seal Harbor (Redfield); — Indian Point road, Somesville (Faxon & Rand).

C. varia, Muhl. *C. Emmonsii,* Dewey.

Infrequent. In dry ground, Sutton Island (Redfield); — Beech Hill cross-road (Rand); — Norwood Road, Southwest Harbor (Faxon & Rand).

C. Novæ-Angliæ, Schw.

Common, especially in dry clearings. Beech Cliff ; Somes Pond pastures; Intervale Brook valley; Sargent District; Seal Cove; Indian Point road, Somesville; Norwood Road, Southwest Harbor; head of The Barcelona, etc. (Rand); — Breakneck road, etc. (Redfield).

C. Pennsylvanica, Lam.

Infrequent. Abundant, however, along the Indian Point road between Somesville and Oak Hill (R. & R.); — Breakneck road, near Hulls Cove (Redfield).

C. communis, Bailey. *C. varia,* Gray, Man., 5th ed.

Common in dry ground everywhere. A form from Beech Mt. Notch "approaching var. *Wheeleri,* Bailey" (Rand). A form from clearing on Meadow Brook, Indian Point road, Somesville, with very soft, short, bright green leaves (Faxon & Rand). Perigynia of this species much infested by a smut.

C. umbellata, Schk.

"Very low and compact, with the spikes all closely clustered near the surface of the ground." L. H. Bailey, Bull. Torr.

Bot. Club, xvi. 219. Frequent. High Head; Browns Mt.; Youngs District, etc. (Rand);—Breakneck road (Redfield).

Var. vicina, Dewey.

"Looser and taller than the type, with many of the peduncles elongated and becoming true culms." Bailey, *l. c.* More common than the type. Sargent Mt.; clearings, near Sunken Heath; Intervale Brook valley; Beech Mt. Notch, etc. (Rand);—Somesville; Sargent Mt.; Beech Cliff (R. & R.). Forms intermediate between this and the type are not infrequent.

C. polytrichoides, Muhl.

Low ground, and damp grassy places; common.

C. stipata, Muhl.

Very common, and variable.

C. tenella, Schk.

Damp places; infrequent. Sargent Mt.; woods on Town Hill road, Somesville (Rand).

C. exilis, Dewey.

Swamps and pond borders; frequent. Breakneck Ponds (R. & R.);—Somes Pond (E. Faxon);—Sunken Heath (Rand). Perigynia much infested by a smut.

C. sterilis, Willd. *C. echinata*, Murray, var. *microstachys*, Boeckl. Gray, Man., 6th ed., 618.

Short, stiff, and erect (usually not much exceeding 1° in height), the old leaves often persistent; head tawny or greenish-yellow, short, composed of from three to five small loosish contiguous spikes, of which the uppermost is usually conspicuously attenuated at the base by the presence of staminate flowers,—sometimes the terminal spike, or even the whole head, is entirely staminate; perigynium thin and flat, conspicuously contracted into a slender beak,—which is nearly or quite as long as the body and spreading so as to give the spike an echinate appearance,—sharp-edged and rough on the upper margins, variously nerved and very sharply toothed. L. H. Bailey, Bull. Torr. Bot. Club, xx. 424. Common in bogs and meadows. Head of

Northeast Creek ; Sargent Mt.; High Head, and elsewhere (Rand); — Somesville; Sea Wall (R. & R.).

Var. excelsior, Bailey.

Taller and more slender (often 2° high), the heads usually more scattered and mostly somewhat greener. Bailey, *l. c.* Common in bogs and low grounds throughout the Island.

Var. cephalantha, Bailey. *C. echinata*, Murray, var. *cephalantha*, Bailey. Gray, Man., 6th ed., 618.

Rather stiff but slender and tall, or the top of the culm weak (1°–2° high); head mostly continuous or more or less dense and composed of five to eight approximate (rarely scattered), large (15–30-flowered) green or greenish loose spikes, in which the mature narrow long-beaked perigynia usually spread nearly or quite at right angles. Bailey, *l. c.* Frequent. Little Cranberry Isle; Barr Hill; Seal Harbor (Redfield) ; — meadow, Doctors Brook; High Head meadow (Rand) ; — Salisbury Cove (R. & R.).

Var. angustata (Carey), Bailey. *C. echinata*, Murray, var. *angustata*, Bailey. Gray, Man., 6th ed., 618.

Very slender, sometimes almost thread-like, weak, bearing long and narrow divaricate perigynia, which are either in loose small heads or in scattered spikes. Bailey, *l. c.* Rare. Wet ground at junction of Prettymarsh and Seal Cove roads (Rand).

C. Atlantica, Bailey. *C. echinata*, Murray, var. *conferta*, Bailey. Gray, Man., 6th ed., 618.

Tall (16′–24′) and very stiff and strong, the leaves broad but stiff and usually becoming somewhat involute when dry; spikes contiguous or scattered, spreading, globular or short-cylindrical, densely flowered, green; the terminal one slenderly contracted below or even entirely staminate; perigynium large and very broad (the body about as broad as long), with a distinct rough, bifid beak, strongly many-nerved, especially upon the back, squarrose or usually retrorse at maturity, shelling off readily when ripe. Bailey, *l. c.* Frequent. Sargent Mt.; Freeman Heath ; meadow on Sunken Heath Brook ; wood road to Aunt Bettys Pond (Rand). None of the specimens thus far collected are really typical.

C. canescens, L.

Wet grounds ; common everywhere. Forms approaching var. *alpicola,* Wahl., occur on Indian Point road, Somesville ; in woods near Spruce Point, Eden ; and near northern end of Denning Pond (Rand).

Var. **vulgaris,** Bailey.

Common as the type ; in woods and drier places.

C. Norvegica, Willd.

Rare. Borders of salt marsh, Little Cranberry Isle (Redfield).

C. trisperma, Dewey.

Common in bogs and wet ground.

C. Deweyana, Schw.

Rare. Southwest Valley road (Greenleaf, Lane & Rand).

C. tribuloides, Wahl.

Northeast Meadow (R. & R.); — burnt woods, Youngs District (Rand). Commonly appearing in the next named variety.

Var. **reducta,** Bailey.

More common than the type. In copse, near bridge at mouth of Northeast Creek; High Head meadow; Oak Hill; bog at northern foot of Beech Hill (Rand).

Var. **cristata** (Schw.), Bailey. *C. cristata,* Schw.

Rare. Damp roadside at northern foot of Beech Hill (Rand).

C. scoparia, Schk.

Common everywhere.

Var. **minor,** Boott.

Dry ground; infrequent. Beech Mt. Notch; Youngs District; Somesville; road west of Browns Mt. (Rand).

C. adusta, Boott. *C. pinguis,* Bailey, not *C. adusta* of Gray, Man., 5th ed.

Local, but not uncommon. Mt. Kebo (Greenleaf); — burnt woods, south of Sunken Heath (Faxon & Rand); — Sea Wall road, Southwest Harbor; Intervale Brook valley; road west of Browns Mt. (Rand).

C. fœnea, Willd.   *C. adusta* of Gray, Man., 5th ed.
Dry ground; common.

Var. **perplexa,** Bailey.
Rare.   Somesville (Redfield); — Beech Hill (Rand).

C. silicea, Olney.
Rare.   Beach near Thumbcap, Great Cranberry Isle (R. & R.).

C. straminea, Willd.   *C. straminea,* var. *tenera* of Gray, Man.,
5th ed.
Infrequent.   Woods, Seal Harbor (Redfield); — meadow on
Denning Brook; Somesville (Rand).

Var. **brevior,** Dewey.
Infrequent.   Mt. Kebo (Greenleaf); — Seal Cove (Rand); —
Town Hill (M. L. Fernald).   A depauperate form, Newport Mt.
(Rand).

Var. **aperta,** Boott.
Common in wet ground, especially near the coast.

Var. **invisa,** W. Boott.
Rare.   Schooner Head (W. Boott, spec. in Herb. Gray); —
bog on shore south of Sea Wall (Rand); — Southwest Harbor
(M. L. Fernald).   A form nearly approaching this variety, Bass
Harbor (Rand).

Var. **alata** (Torr.), Bailey.   *C. alata,* Torr.
Seal Harbor (Redfield).   Specimens not entirely characteristic, but apparently this variety.

C. albolutescens, Schw.   *C. straminea,* Willd., var. *fœnea,* Torr.
Gray, Man., 6th ed., 622.*
Infrequent.   Road west of Browns Mt. (Rand), and probably
elsewhere.

Var. **cumulata,** Bailey.   *C. straminea,* Willd., var. *cumulata,*
Bailey.   Gray, Man., 6th ed., 622.
Dry ground, especially in newly disturbed soil; common.

* See also Bailey in Bull. Torr. Bot. Club, xx. 421, 422.

ANTHOXANTHUM, L.   Sweet Vernal Grass.

A. odoratum, L.

Fields and meadows; becoming common.   Naturalized from Europe.

HIEROCHLOE, Gmelin.   Sweet Grass.

H. borealis, Roem. & Schultes.

Borders of salt or brackish meadows and marshes.   Bass Harbor; Southwest Harbor; Little Harbor; Seal Harbor; Northeast Meadow; Thomas Bay; Cranberry Isles; Duck Islands, and elsewhere.   Much used by the Indians for basket work.

STIPA, L.   Feather Grass.

S. Macounii, Scribner.*   *S. Richardsonii*, Gray, *non* Link.   Gray, Man., 6th ed., 641.

Rare.   Burnt woods, Youngs District (Rand, E. Faxon); — wood clearings, Somesville (M. L. Fernald).

ORYZOPSIS, Mx.   Mountain Rice.

O. asperifolia, Mx.

Dry woods and clearings; not uncommon.

MUHLENBERGIA, Schreb.   Drop-seed Grass.

M. glomerata (Willd.), Trin.

Infrequent.   Damp field, Northeast Harbor (Rand); — Long Pond meadows (Redfield); — Sargent Mt.; Beech Mt. Notch; shore, Jordan Pond (Rand).

* M. diffusa, Schreb.

"In abundance," western slope of Green Mt. (Arnold Greene).

BRACHYELYTRUM, Beauv.

B. aristatum (Pers.), Beauv.

Low woods throughout the Island; frequent, but not abundant.

* Bull. Torr. Bot. Club, xix. 154.

PHLEUM, L. CAT'S-TAIL GRASS.

P. PRATENSE, L. TIMOTHY.

Common in fields and meadows. Naturalized from Europe.

ALOPECURUS, L. FOX-TAIL GRASS.

A. PRATENSIS, L. MEADOW FOX-TAIL.

Fields and by roadsides; infrequent. Seal Harbor (Redfield); — Southwest Harbor ; Bar Harbor ; Hulls Cove; Somesville (Rand). Naturalized from Europe.

SPOROBOLUS, R. Br. DROP-SEED GRASS.

S. serotinus (Torr.), Gray. MIST GRASS.

Meadows and low ground; frequent.

AGROSTIS, L. BENT GRASS.

A. ALBA, L. WHITE BENT GRASS. HERD'S GRASS.

Meadows and fields; common. Naturalized from Europe through cultivation.

Var. SYLVATICA (L.), Scribner.

A viviparous form of the species. Beech Hill; Somes Stream (Rand).

Var. coarctata (Hoffm.), Scribner.

Panicle dense; branches short, flower-bearing to the base; plant not stoloniferous. In wet ground ; frequent. Baker Island (Redfield); — High Head; bog, Kings Point, Southwest Harbor (Rand); — head of Somes Sound (Greenleaf). Doubt-less indigenous.

Var. stolonifera (L.), Vasey.

Panicle dense, but the spikelets less crowded than in var. *coarctata*, narrow, often linear; plant stoloniferous. On beaches by the shore, and elsewhere. Hunters Beach; High Head (Rand) ; — Little Cranberry Isle; Little Harbor (Redfield). Doubtless indigenous.

ANTHOXANTHUM L. Sweet Vernal Grass

A. odoratum, L.

Fields and meadows; becoming common. Introduced from Europe.

HIEROCHLOE Gmelin. Holy Grass

H. borealis, Room. & Schultes

Borders of salt or brackish meadows and marshes. Bar Harbor, Southwest Harbor, Little Harbor and Harbor, Northeast Meadow, Thomas Bay, Cranberry Isles, Duck Islands, and elsewhere. Much used by the Indians for basket work.

STIPA L. Feather Grass

S. Macounii, Scribner.* *S. Richardsonii* var. non Link. Gray, Man., 6th ed. 641.

Rare. Burnt woods, Thomas District, Bass & Femald,—wood clearings, Somesville M. L. Femald.

ORYZOPSIS Mx. Mountain Rice

O. asperifolia. Mx.

Dry woods and clearings; not uncommon.

MUHLENBERGIA Schreb. Drop-seed Grass

M. glomerata (Willd.), Trin.

Infrequent. Damp field, Northeast Harbor Rand);—Long Pond meadows (Redfield);—Sargent Mt.; Beech Mt. North shore, Jordan Pond (Rand).

* **M. diffusa,** Schreb.

"In abundance." western slope of Green Mt. (Arnold Greene).

BRACHYELYTRUM Beauv.

B. aristatum (Pers.), Beauv.

Low woods throughout the Island; frequent, but not abundant.

PHLEUM, L. CAT'S-TAIL GRASS.

P. PRATENSE, L. TIMOTHY.

Common in fields and meadows. Naturalized from Europe.

ALOPECURUS, L. FOX-TAIL GRASS.

A. PRATENSIS, L. MEADOW FOX-TAIL.

Fields and by roadsides; infrequent. Seal Harbor (Redfield); — Southwest Harbor ; Bar Harbor ; Hulls Cove; Somesville (Rand). Naturalized from Europe.

SPOROBOLUS, R. Br. DROP-SEED GRASS.

S. serotinus (Torr.), Gray. MIST GRASS.

Meadows and low ground; frequent.

AGROSTIS, L. BENT GRASS.

A. ALBA, L. WHITE BENT GRASS. HERD'S GRASS.

Meadows and fields; common. Naturalized from Europe through cultivation.

Var. SYLVATICA (L.), Scribner.

A viviparous form of the species. Beech Hill; Somes Stream (Rand).

Var. coarctata (Hoffm.), Scribner.

Panicle dense; branches short, flower-bearing to the base; plant not stoloniferous. In wet ground ; frequent. Baker Island (Redfield); — High Head; bog, Kings Point, Southwest Harbor (Rand); — head of Somes Sound (Greenleaf). Doubtless indigenous.

Var. stolonifera (L.), Vasey.

Panicle dense, but the spikelets less crowded than in var. *coarctata*, narrow, often linear; plant stoloniferous. On beaches by the shore, and elsewhere. Hunters Beach; High Head (Rand) ; — Little Cranberry Isle; Little Harbor (Redfield). Doubtless indigenous.

Var. VULGARIS (With.), Thurb. RED-TOP.

Meadows and fields; common. Naturalized from Europe through cultivation, and perhaps also indigenous.

A. perennans (Walt.), Tuck.

Frequent in damp shady places and by brooksides. Beech Mt. Notch; Deer Brook, Jordan Pond; Gilmore Brook; Little Harbor Brook Notch, and elsewhere (Rand);—Southwest Harbor (M. L. Fernald). It seems doubtful whether these northern plants belong to the true southern *A. perennans*. A form from woods south of Beech Mt. may be *A. Novæ-Angliæ*, Tuck.

A. scabra, Willd. HAIR GRASS.

Dry soil; common. A dwarf form growing in tufts in rock hollows and dry places is var. *montanum* (Torr.). Sargent Mt. (Greenleaf, Rand); — Western Mt.; White Beach, Great Pond (Rand).

A. canina, L.

Infrequent. Sargent Mt. (Greenleaf, Rand).

CINNA, L. WOOD REED GRASS.

C. pendula, Trin.

Damp woods. Cold Brook ; Little Harbor Brook valley; Beech Mt. Notch (Rand); — Seal Harbor (Redfield). A very robust form, head of Beech Mt. Notch (Rand).

CALAMAGROSTIS, Adans. REED BENT GRASS.

C. Canadensis (Mx.), Beauv. BLUE JOINT.

Common in moist or wet ground from sea level to mountain summits.

DESCHAMPSIA, Beauv. HAIR GRASS.

D. flexuosa (L.), Trin.

Common in dry places, especially on hills and mountains.

TRISETUM, Pers.

T. subspicatum (L.), Beauv., var. **molle** (Mx.), Gray. Rare. Bluff by shore, Northwest Cove (Rand).

DANTHONIA, DC. Wild Oat Grass.

D. spicata (L.), Beauv.
Common everywhere in poor soil.

DACTYLIS, L. Orchard Grass.

D. glomerata, L.
Sparingly introduced. Northeast Harbor; "Fox Dens," Southwest Harbor; Somesville (Rand). Naturalized from Europe.

POA, L. Spear Grass. Meadow Grass.

P. annua, L. Low Spear Grass.
Roadsides and cultivated grounds. Bar Harbor; Southwest Harbor; Northeast Harbor; Somesville, and elsewhere. Introduced, and becoming common. Naturalized from Europe.

P. compressa, L. Wire Grass.
Frequent in dry soil, or rocky places. Appearing indigenous in many places on the Island, but doubtless naturalized from Europe.

P. nemoralis, L.
Somesville (Rand).

P. serotina, Ehrh. False Red-top.
Common and very variable. This species runs gradually into *P. nemoralis*; specimens from Bubble Pond (Rand) can hardly be distinguished. A large form with ample panicle (*P. fertilis*, Host), Somes Pond, at outlet, in water (Rand).

P. pratensis, L. Kentucky Blue Grass.
Common. Perhaps indigenous, but mostly naturalized from Europe.

GLYCERIA, R. Br. MANNA GRASS.

G. Canadensis (Mx.), Trin. RATTLESNAKE GRASS. JOB'S TEARS.
Common in wet places.

G. laxa, Scribner.

A coarse leafy grass, 2°–4° high, with a diffuse ample panicle
and oblong, somewhat turgid spikelets. Sheaths scabrous, the
lower exceeding the internodes. Ligule about 1″ long, thin,
lacerate. Leaves 8′–15′ long, 3″–4″ wide, very rough-scabrous,
both sides tapering to a sharp point or the lower ones abruptly
sharp-pointed. Panicle 7′–9′ long, the main axis and branches
strongly scabrous, lower branches in twos or threes, the upper
solitary, the longer and usually widely spreading lower ones
3′–5′ long. Spikelets oblong or broadly ovate, 3–5-flowered,
about 2″ long, much exceeding the pedicels, and from 1″–1½″
wide. Empty glumes unequal, scarious-margined, the larger
second glume about one half the length of the first floret.
Flowering glumes rounded on the back, 1″–1¼″ long, broadly
obovate, obtuse, with a narrow scarious margin above, 7-nerved,
nerves not prominent. Palea nearly equalling the glume, the
keels smooth, strongly curved above. Closely allied to *G. Canadensis,* but the smaller spikelets are green or purple tinged, and
the more obtuse floral glume scarcely exceeds the narrower palea.
F. Lamson-Scribner in Bull. Torr. Bot. Club, xxi. 37, *sub nom.*
Panicularia laxa, and republished here as above at his desire.
Seal Harbor (Redfield) ; — Somesville (Rand). A form apparently of this species with smaller and imperfectly developed
spikelets, Great Cranberry Isle; Seal Harbor (Redfield).

G. obtusa (Muhl.), Trin.

Common in wet grounds about Somesville. Also Southwest
Harbor; Sea Wall (Rand).

G. elongata (Torr.), Trin.

Wet woods. Near Beech Hill (Arnold Greene) ; — Canada
Valley; Seal Harbor; Beech Mt. Notch (Rand).

G. nervata (Willd.), Trin.

Common in meadows.

G. pallida (Torr.), Trin.

A narrow-leaved form in bog at northern foot of Beech Hill (Rand).

G. grandis, S. Watson. *G. aquatica* (L.), Sm.

Somes Stream, Somesville (Rand).

G. fluitans (L.), R. Br.

Brooksides and ditches; frequent.

PUCCINELLIA, Parl.* (*Atropis*, Rupr.)

P. maritima (Huds.), Parl. *Atropis maritima* (Huds.), Griseb. *Glyceria maritima* (Huds.), Wahl. Sea Spear Grass.

Rare. Shore, Somes Harbor (Redfield). A puzzling form, perhaps a hybrid between this species and *P. distans*, Norwood Cove (M. L. Fernald).

Var. (?) **minor,** S. Watson. *Atropis maritima,* var. *Nutkaensis* (Presl), Scribner. *Atropis angustata* (R. Br.), Griseb. *Puccinellia angustata* (R. Br.). *Glyceria angustata* (R. Br.), Fries.

Sea beaches. Somesville; Ovens; Thomas Bay; Southwest Harbor; Mt. Desert Narrows, and elsewhere. "This form is distinguished from *P. maritima* by its smaller and weaker habit, and by having the keels of the palea smooth below and only very minutely scabrous above. In *P. maritima* the keels of the palea are strongly fringed nearly or quite to the base." F. Lamson-Scribner. More recent examination shows this form to be specifically distinct, and should bear the name in the Manual of *P. angustata* (R. Br.).

FESTUCA, L. Fescue Grass.

F. ovina, L. Sheep's Fescue.

Common especially on or near the shore. Also on Cranberry Isles. All specimens from Mt. Desert and vicinity are referred, however, to *F. rubra*, L., by Prof. F. Lamson-Scribner, which

---

* Better considered as a sub-genus of Glyceria.

he considers without doubt a distinct species from *F. ovina.* Specimens from High Head, Great Cranberry Isle, Little Cranberry Isle, and Thompson Island, may be referred to var. *genuina,* Hack. A specimen from Indian Point road, Somesville, appears to be var. *fallax,* Hack.

"In *F. rubra* the leaves of the culm and sterile shoots are similar, the ligules in the latter are not auriculate, and the shoots themselves are extra-vaginal; i. e. the buds of the branches at the base of the culm burst through the base of the leaf sheath in the axil of which they are formed. In *F. ovina* the leaves of the flowering culms and sterile shoots are unlike, the ligules on the latter are auriculate and the shoots themselves are intra-vaginal; i. e. the buds in the lower leaf axils grow up out of the sheaths and do not break through them below. *F. ovina* is strictly tufted, while *F. rubra* extends more or less by rootstocks." F. Lamson-Scribner.

F. DURIUSCULA, L. *F. ovina,* L., var. *duriuscula,* Koch. Gray, Man., 6th ed.

Rare. Near Hulls Cove (Rand). Naturalized from Europe.

F. ELATIOR, L.

Fields and roadsides; common. Northeast Harbor; Southwest Harbor; Bar Harbor; Somesville; Seal Harbor; High Head. Naturalized from Europe.

Var. PRATENSIS (Huds.), Gray.

Fields. Seal Harbor; Southwest Harbor; Somesville; High Head, and elsewhere. Naturalized from Europe.

BROMUS, L.   BROME GRASS.

B. ciliatus, L.

Frequent in rocky woods and low ground. Variable.

AGROPYRUM, Gærtn.   (*Triticum,* L.)   FALSE WHEAT.

A. repens (L.), Beauv. *Triticum repens,* L.   QUITCH GRASS. WITCH GRASS.

Fields, waysides, and shores; common and very variable. Naturalized from Europe in cultivated grounds, and also indigenous. The Island forms appear for the most part to be indigenous northern and coast forms, and abound on beaches and on banks by the shore.

Var. **glaucum** (Desf.), Boiss. *Triticum repens*, L., var. *intermedium*, Fries.

A glaucous, rigid, maritime form, with large crowded spikelets and glumes blunt or mucronate. Seal Harbor (Redfield); — Northeast Harbor; Southwest Harbor, etc. (Rand).

Var. **pilosum**, Scribner.

Upper surface of leaves pilose, rhachis of spike pubescent to hirsute, flowering glumes awnless or short cuspidate pointed. Southwest Harbor (Rand). This, however, may be the same as var. *agreste*, Anders.

Other well marked forms are numerous; but it seems impossible to identify them with any certainty without a careful comparison with authentic specimens in European herbaria.

A. caninum (L.), Roem. & Schultes. *Triticum caninum*, L.

Rare. Field near the head of Northeast Harbor (Rand). Naturalized from Europe, and also indigenous. Probably introduced on the Island.

HORDEUM, L. BARLEY.

H. jubatum, L. SQUIRREL-TAIL GRASS.

Common on the coast.

ELYMUS, L. WILD RYE. LYME GRASS.

E. Virginicus, L.

Common on the coast.

E. mollis, Trin.

Muddy or sandy shores on the coast; frequent. Considered by many authors as identical with *E. arenarius*, L., which occurs on the Pacific coast.

SERIES II. CRYPTOGAMIA; FLOWERLESS PLANTS.

CLASS I. PTERIDOPHYTA.

EQUISETACEÆ. HORSETAIL FAMILY.

EQUISETUM, L. HORSETAIL. SCOURING RUSH.

E. arvense, L. COMMON HORSETAIL.
Moist gravelly soil; common.

E. sylvaticum, L.
Wet banks and shady places; frequent.

E. limosum, L.
Borders of ponds; rare. Bubble Pond (F. M. Day, Redfield);
— Great Pond (Rand).

FILICES. FERNS.

POLYPODIUM, L. POLYPODY.

P. vulgare, L.
Rocks; very common.

PTERIS, L. BRAKE.

P. aquilina, L.
Dry soil; very common.

ASPLENIUM, L. SPLEENWORT.

A. Filix-fœmina (L.), Bernh. LADY FERN.
Damp shady places; frequent, and very variable.

Var. angustatum (Willd.), D. C. Eaton. Var. *Michauxii*,
Mett.

Fronds 1°–3° high, rather rigid, narrow in outline, nearly
bipinnate; pinnæ obliquely ascending or curved upwards, narrowly lanceolate; segments oblong, crowded, crenated or serrate;

sori usually abundant, straight or curved. Eaton, Ferns N. A. 227. Woods; infrequent. Salisbury Cove (Clara L. Walley).

Var. exile, D. C. Eaton.

Fronds 3'-6' high, lanceolate, pinnate; pinnæ oblong-lanceolate, deeply cut into oblong laciniæ which are two- to three-toothed at the end. Eaton, Ferns N. A. 227. Woods; rare. Breakneck road (Clara L. Walley).

PHEGOPTERIS, Fée. BEECH FERN.

P. polypodioides, Fée. *Polypodium Phegopteris, L.*

Common in damp woods. A form closely approaching *P. hexagonoptera*, Fée, in woods, Seal Harbor (Redfield). A form with the main rhachis forked at the apex, woods, head of The Barcelona meadow (Rand).

P. Dryopteris (L.), Fée.

Damp woods and shaded rocky places; common.

ASPIDIUM, Swz. SHIELD FERN.

A. Thelypteris (L.), Swz.

Damp ground; not uncommon. Seal Harbor; Long Pond meadows; Sutton Island; Great Cranberry Isle; Baker Island (Redfield); — on Somes Stream (R. & R.).

A. Noveboracense (L.), Swz.

Damp woods; frequent. Seal Harbor; Great Cranberry Isle, etc. (Redfield); — Breakneck road (Clara L. Walley); — "Mt. Desert" (F. M. Day).

A. spinulosum (Retz), Swz.

Damp woods; not uncommon. Breakneck road (Clara L. Walley); — Seal Harbor; Ovens (Redfield).

Var. intermedium (Muhl.), D. C. Eaton.

Woods; common.

Var. dilatatum (Hoffm.), Hook.

Woods; infrequent. Breakneck road (Clara L. Walley); — west of Southwest Harbor (M. L. Fernald). A dwarf form in woods, Breakneck road (Clara L. Walley).

A. cristatum (L.), Swz.

Wet ground; frequent.

Var. Clintonianum, D. C. Eaton.

Rare. Near Breakneck road (Clara L. Walley).

A. marginale (L.), Swz.

Rocky woods; frequent.

A. acrostichoides (Mx.), Swz. CHRISTMAS FERN.

Deep rocky woods; frequent.

CYSTOPTERIS, Bernh. BLADDER FERN.

C. fragilis (L.), Bernh.

Rare. Caves, Barr Hill (Redfield); — wet cliffs, West Branch of Hadlock Brook (Rand).

ONOCLEA, L. SENSITIVE FERN.

O. sensibilis, L.

Common in low ground.

WOODSIA, R. Br.

W. Ilvensis (L.), R. Br.

Infrequent and local. Dog Mt. (Rand); — Flying Mt. (Annie S. Downs, H. C. Jones, and others); — Beech Cliff (Annie S. Downs, R. & R.).

DICKSONIA, L'Hér.

D. pilosiuscula, Willd. *D. punctilobula* (Mx.), Gray.

Common everywhere.

OSMUNDA, L. Flowering Fern.

O. regalis, L. Flowering Fern.

Swamps and wet meadows; common.

O. Claytoniana, L. Interrupted Flowering Fern.

Common in low ground. A form with upper pinnæ of sterile frond partly fertile, Bar Harbor (Mary Minot). A form with middle pinnæ of fertile frond partly sterile, and undeveloped, Seal Harbor (Redfield).

O. cinnamomea, L. Cinnamon Fern.

Common in damp ground everywhere.

OPHIOGLOSSACEÆ. Adder's Tongue Family.

BOTRYCHIUM, Swz. Moonwort.

B. simplex, Hitchcock.

Rare. High Head (Rand, Redfield & Faxon); — Beech Cliff (Rand).

B. matricariæfolium, A. Br.

Infrequent. Jordan Pond road (Harriet A. Hill); — field, Northwest Cove; Cold Brook; Great Cranberry Isle (Rand).

B. ternatum (Thunb.), Swz.

Somewhat frequent in low fields and pastures; occasionally in woods. Seal Harbor; Long Pond meadows (Redfield); — Sutton Island; Sawyer Cove (Harriet A. Hill); — Duck Brook road (Clara L. Walley); — woods, north of Jordan Pond (Theodore G. White); — meadow, head of Northeast Creek; Southwest Valley road (Rand).

Var. **intermedium,** D. C. Eaton.

Rare. High Head meadow (Rand).

Var. **obliquum** (Muhl.), Milde.

Not uncommon. Duck Brook road (Clara L. Walley); — meadow, head of Northeast Creek; Southwest Harbor (Rand); fields above Long Pond (R. & R.).

188 FLORA OF MOUNT DESERT.

Var. dissectum (Spreng.), Milde.

Frequent. Duck Brook road (Clara L. Walley); — meadow, head of Northeast Creek; High Head meadow; fields, Clark Point, Southwest Harbor (Rand); — fields above Long Pond (Redfield); — Sawyer Cove (Harriet A. Hill).

OPHIOGLOSSUM, L. ADDER'S TONGUE.

O. vulgatum, L.

Rare. Wet field, head of Southwest Harbor (Annie S. Downs).

LYCOPODIACEÆ. CLUB MOSS FAMILY.

LYCOPODIUM, L. CLUB MOSS.

L. Selago, L.

Rare. Sargent Mt. (H. C. Jones, Rand, Redfield).

L. lucidulum, Mx.

Deep moist woods; common.

L. inundatum, L.

Bogs and wet places; frequent. Cedar Swamp; Southwest Harbor; Sea Wall Swamp; Aunt Bettys Pond; Ripples Pond; Gilmore Meadow; Great Cranberry Isle. etc. (Rand); — bog, Hadlock Upper Pond, etc. (Redfield).

Var. Bigelovii, Tuck.

Bogs; uncommon. Border of Upper Breakneck Pond; border of Aunt Bettys Pond (Rand).

L. annotinum, L.

Woods and damp thickets ; frequent. Also Sargent Mt. (Theodore G. White).

L. obscurum, L.

Rare. Roadside between Somesville and Southwest Harbor, near Canada Valley (Rand).

Var. dendroideum (Mx.), D. C. Eaton. GROUND PINE.

Woods; common.

L. clavatum, L.

Woods and thickets; very common.

L. complanatum, L. TRAILING CHRISTMAS GREEN.

Dry woods and thickets; common.

Var. **Chamæcyparissus** (Braun), D. C. Eaton.

Woods and mountain thickets; infrequent. Between Jordan Pond and Eagle Lake (Redfield); — between Jordan Pond and Northeast Harbor (R. & R.); — Pemetic Mt.; woods, Aunt Bettys Pond (Rand).

SELAGINELLACEÆ.

SELAGINELLA, Beauv.

S. rupestris (L.), Spring.

Rare and local. Flying Mt. (H. C. Jones, Rand); — Dog Mt. (Elizabeth G. Britton, Rand, E. Faxon).

ISOETES, L. QUILLWORT.

I. lacustris, L.

Infrequent. Mouth of Deer Brook, Jordan Pond (Redfield); — west shore of Jordan Pond (Rand). A very small form on west shore of Jordan Pond (Rand). A peculiar form, probably of this species, Deer Brook (Rand); — Northwest Arm, Great Pond (Fernald).

I. echinospora, Durieu, var. **Braunii** (Durieu), Engelm.

Sandy and gravelly brook courses and pond shores; frequent. Jordan Stream; Deer Brook; shores of Jordan Pond (R. & R.); — Somes Stream; Denning Brook (Rand).

I. riparia, Engelm.

Rare. Somes Stream (George G. Kennedy); — southeast end of Ripples Pond (Rand). A form "not typical *I. riparia,* and yet quite a departure from *I. lacustris,*" — Ripples Brook (Rand, M. L. Fernald), *fide* L. M. Underwood.

Class II. BRYOPHYTA.

Division I. MUSCI; MOSSES.

Order I. SPHAGNACEÆ. Peat Mosses.

List prepared by Edward L. Rand, assisted greatly by Edwin Faxon and Prof. Daniel C. Eaton, and arranged mainly in accordance with the writings of Dr. Carl Warnstorf, of Neuruppin, Germany.

The plants catalogued in the following list have been collected mostly by Edwin Faxon and Edward L. Rand ; the determinations are by Dr. Carl Warnstorf. In view of the great difficulty of finding descriptions of the various species and varieties, it has been thought wise to give freely references to Lesquereux and James's "Mosses of North America," and to Dr. Warnstorf's articles on North American Sphagna, to be found in Coulter's Botanical Gazette, Vol. XV., in the numbers for the months of June, August, September, and October, 1890,— both of which works can be consulted with little trouble. In cases, furthermore, where descriptions are not there given, they have been either translated or specially prepared for this list by Prof. Eaton and Mr. Faxon. It has not seemed best to include *forms* and *sub-forms* herein, since most of them have little value except for the critical student. Very many of them are, however, represented in the Mt. Desert Herbarium. On the other hand, all varieties are given, without regard to the distinctions on which they are founded. The value, however, of most of the so-called color varieties is very doubtful, since careful observations seem to prove beyond question that the color of Sphagnum varies greatly with the season. It is even a matter of doubt whether a given plant of any species may not quickly vary through the influence of temporary external conditions, assuming at one time the form of one variety, at another time the form of another. This doubtful value of varieties of Sphagnum furnishes an additional reason for the exclusion of mere *forms* and *sub-forms* from a local catalogue of plants.

SPHAGNUM, L.   PEAT MOSS.

§ 1.   ACUTIFOLIA.

S. fimbriatum, Wils.   L. & J., Mosses N. A., 14.   Var. tenue. Gravet.   Bot. Gaz., xv. 128.

Swamp west of Sea Wall; Red Rocks, Great Cranberry Isle; Bass Harbor (Rand); — Little Cranberry Isle (Redfield); — Sargent Mt. (E. Faxon).

S. Girgensohnii, Russ.   *S. strictum*, Lindb.   L. & J., Mosses N. A., 13.

Common.

Var. **stachyodes**, Russ.   Bot. Gaz., xv. 129.

Southwest Harbor; western side of Browns Mt.; Seal Harbor; Long Pond meadows (Rand).

Var. **hydrophilum**, Russ.

Plant 5–8 cm. high, pale green, mostly drepanocladous, growing in wet places, the coma indistinct; stem leaves narrow, nearly twice longer than broad, very narrowly margined except at the very base, hyaline cells often partitioned, pores and fibrils none; branch leaves loosely imbricate with spreading tips.   High Head (Rand).

Var. **teretiusculum**, Warnst.

In extremely compact tufts, 5–7 cm. high; plants very slender; stem leaves very small, lingulate, about one and one half times as long as wide; branch leaves also small, closely imbricated, so that the short branches are perfectly terete.   Warnst., Hedwigia, xxxii. (1893) 15.   Summit of Sargent Mt. (Rand).

Var. **sphærocephalum**, Warnst.

In compact tufts about 10 cm. high; stem leaves small, lingulate, little longer than broad; branches of the coma united into a large, thick, almost spherical head, with leaves remarkably large and in part squarrosely spreading.   Warnst., Hedwigia, xxxii. (1893) 15.   Reservoir, Jordan Pond road (Rand).

S. Russowii, Warnst. Bot. Gaz., xv. 130.

Pemetic Mt. (Rand).

Var. **pœcilum,** Russ. Bot. Gaz., xv. 132.

Beech Mt. (Rand).

Var. **rhodochroum,** Russ. Bot. Gaz., xv. 132.

Breakneck Ponds (E. Faxon).

Var. **carneum,** Russ.

Upper part of plant pale flesh-color, passing below into pale gray or grayish green. Woods, Norwood Cove; woods, southern foot of Dog Mt. (Rand).

S. fuscum (Schimp.), von Klinggraef. Bot. Gaz., xv. 133. *S. acutifolium,* var. *fuscum,* Schimp. L. & J., Mosses N. A., 13. Var. **fuscescens,** Warnst. Bot. Gaz., xv. 135.

Common in dry bogs. Freeman Heath; Sunken Heath; Aunt Bettys Pond (Faxon & Rand); — The Heath, Great Cranberry Isle; Sea Wall, etc. (Rand).

S. tenellum (Schimp.), von Klinggraef. Bot. Gaz., xv. 135.

Browns Mt.; woods, Sea Wall Swamp (E. Faxon).

Var. **rubellum** (Wils.), von Klinggraef. Bot. Gaz., xv. 137. *S. rubellum,* Wils. L. & J., Mosses N. A., 13.

Common in various forms.

Var. **versicolor,** Warnst. Bot. Gaz., xv. 137.

Pond Heath; Sunken Heath (E. Faxon).

Var. **violascens,** Warnst.

Color above, a livid mixture of violet, red, and green; below, pale reddish. Lower Breakneck Pond (E. Faxon).

S. Warnstorfii, Russ. Bot. Gaz., xv. 138. Var. **violascens,** Warnst.

Plants pale yellowish green mixed with pale violet or violet-red. High Head meadow (E. Faxon).

Var. **purpurascens,** Russ. Bot. Gaz., xv. 140.

High Head meadow (Rand).

S. acutifolium (Ehrh. in part), Russ. & Warnst. Bot. Gaz., xv. 191.

Common.

Var. rubrum (Brid.), Warnst.

Plants rosy-red above, gradually becoming paler below. Common.

Var. versicolor, Warnst. Bot. Gaz., xv. 193.

Frequent. Freeman Heath ; Beech Hill (E. Faxon); — Breakneck Ponds; Sargent Mt.; Beech Mt. Notch (Rand).

Var. viride, Warnst. Bot. Gaz., xv. 193.

Beech Mt.; Cold Brook; Southwest Harbor (Rand).

Var. pallescens, Warnst. Bot. Gaz., xv. 193.

Sargent Mt.; Beech Mt. Notch (Rand).

S. subnitens, Russ. & Warnst., var. **flavicomans**, Card. Bot. Gaz., xv. 194–196.

Frequent. Swamp, Meadow Brook, Somesville; Breakneck Ponds; Freeman Heath; Sea Wall Swamp (E. Faxon); — bog west of Hio; Great Cranberry Isle; Somes Pond, etc. (Rand).

Var. obscurum, Warnst. Bot. Gaz., xv. 196.

Little Cranberry Isle (Redfield).

Var. violascens, Warnst.

Whole plant very pale, soft; white with faint tinge of violet; coma faint pink. Breakneck Ponds (Rand).

Var. pallescens, Warnst.

Plant soft, whitish, becoming pale green toward the coma, but without any tinge of brownish yellow. Breakneck Ponds (Faxon & Rand); — bog west of Hio (Rand).

§ 2. CUSPIDATA.

S. recurvum (Beauv.), Russ. & Warnst. *S. intermedium*, Hoffm. L. & J., Mosses N. A., 15.

Sea Wall Swamp; Sargent Mt. (E. Faxon); — Breakneck Ponds (Rand).

Var. pulchrum, Lindb. Bot. Gaz., xv. 218.

Frequent; found in many forms, of which *forma fuscescens*, with the whole plant rich golden brown in color, is the most beautiful. Abundant at Sunken Heath ; Breakneck Ponds (Faxon & Rand); — Sea Wall; Great Cranberry Isle (Rand).

Var. mucronatum, Russ. Bot. Gaz., xv. 218.

Breakneck Ponds; Otter Cliffs (Rand).

Var. amblyphyllum, Russ. Bot. Gaz., xv. 219.

Northwest Arm woods (Rand); — near Northwest Cove; Sea Wall Swamp (E. Faxon).

Var. parvifolium (Sendt.), Warnst. Bot. Gaz., xv. 219.

Pond Heath (E. Faxon); — bog near Sea Wall (Rand).

S. cuspidatum (Ehrh.), Russ. & Warnst., var. Miquelonense, Ren. & Card. Bot. Gaz., xv. 230.

Frequent. Round Pond; near Northwest Cove (E. Faxon); — Aunt Bettys Pond; Northwest Arm woods, etc. (Faxon & Rand); — Great Cranberry Isle (Rand).

Var. falcatum, Russ. L. & J., Mosses N. A., 15.

Branches distinctly falcate at the apex. Pools, Great Heath; The Heath, Great Cranberry Isle (Rand).

Var. submersum, Schimp.

Tufts loose, very soft, deep green, nearly or quite submersed or floating; stem very long and slender, green; branches rather long, decurved; stem leaves broadly ovate-oblong, pointed, fibrillose near the apex; branch leaves rather long and narrow, green, flexuous when dry; fruit scattered along the stem below the coma, pseudopodia often very long, perichætial leaves scattered, fibrillose. Plant softer than var. *Miquelonense,* and much smaller than var. *Torreyanum.* Frequent. In pools, Freeman Heath; Dog Mt. (Faxon); — Beech Mt. (Faxon & Rand); — Sea Wall; Sunken Heath (Rand);—Great Cranberry Isle (Redfield).

Var. plumulosum, Schimp. L. & J., Mosses N. A., 15.

Tufts soft, compact; stems short and branches erect; leaves short, lanceolate-subulate, very narrow. Pools, Sunken Heath

(Faxon & Rand); — The Heath, Great Cranberry Isle; Great Heath (Rand).

S. Dusenii (Jens.), Russ. & Warnst. *S. Mendocinum,* Warnst. in Bot. Gaz., xv. 221, not of S. & L. Var. **parvifolium,** Warnst.

Plants soft and slender, in dense tufts, partly immersed; stem leaves small, about 0.54–0.60 mm. long and as broad at base, triangular-lingulate, without fibrils or somewhat fibrillose toward the usually rounded and slightly fimbriate apex; branch leaves also small, about 1.14–1.37 mm. long and 0.54 mm. broad, almost always falcate-secund, narrowly bordered; outer pores numerous, often passing into large membrane-gaps toward the apex. Warnst., Hedwigia, xxxii. (1893) 14. In pool, Dog Mt. (E. Faxon).

S. molluscum, Bruch. *S. tenellum,* Ehrh. L. & J., Mosses N. A., 20.

Abundant, Sunken Heath (Faxon & Rand); — Great Heath (E. Faxon); — The Heath, Great Cranberry Isle (Rand). The more robust forms are var. *robustum,* Warnst.

§ 3. SQUARROSA.

S. squarrosum, Pers. L. & J., Mosses N. A., 16.

Frequent. Near Northwest Cove (Faxon); — Seal Harbor; woods, Somes Pond (Rand).

Var. **spectabile,** Russ. Bot. Gaz., xv. 224.

Not uncommon.

Var. **semisquarrosum,** Russ. Bot. Gaz., xv. 224.

Southwest Valley road (Rand).

S. teres, Angstr. L. & J., Mosses N. A., 16.

Rare. The Barcelona meadow (E. Faxon).

Var. **imbricatum,** Warnst. Bot. Gaz., xv. 224.

Bog, north of Beech Hill (Rand).

§ 4. POLYCLADA.

S. Wulfianum, Girgens.   L. & J., Mosses N. A., 16.   Var. **viride,** Warnst.   Bot. Gaz., xv. 225.

Rare.   Cold Brook; Beech Mt. (Rand).

§ 5. RIGIDA.

S. compactum, DC.   *S. rigidum,* Schimp.   L. & J., Mosses N. A., 17.

Frequent in its three varieties.

Var. **squarrosum,** Russ.   Bot. Gaz., xv. 226.

Woods, Lower Breakneck Pond (Faxon & Rand); — Southwest Harbor; Seal Harbor; Sargent Mt. (Rand); — Robinson Mt. (Faxon).

Var. **subsquarrosum,** Warnst.   Bot. Gaz., xv. 226.

Sargent Mt. (Faxon); — Lower Breakneck Pond; Browns Mt.; Dog Mt.; Southwest Harbor; Great Cranberry Isle (Rand).

Var. **imbricatum,** Warnst.   Bot. Gaz., xv. 226.

Sargent Mt. (Faxon & Rand); — Dog Mt. (Rand).

S. Garberi, L. & J.   Mosses N. A., 18.   Var. **squarrosulum,** Warnst.

Tufts pale or bluish-green, low or even 20 cm. high, and then quite similar in habit to *S. compactum,* var. *squarrosulum ;* branch leaves all with the apical half squarrose-recurved. Warnst., Hedwigia, xxxii. (1893) 15.   Very rare.   Sargent Mt. (Rand).   The specimens are *forma sphœrocephalum,* Warnst., which is extremely robust, with the comal branches gathered into a large globular head.   This form is exceedingly rare, there being only one other known station.

Var. **subsquarrosum,** Warnst.

Branch leaves generally merely curved outward, only here and there squarrose.   Warnst., Hedwigia, xxxii. (1893) 15.   Rare. Beech Mt. (Rand & Faxon); — Sargent Mt. (Rand).

§ 6. SUBSECUNDA.

S. Pylæsii, Brid. L. & J., Mosses N. A., 23. Var. **ramosum,** Warnst. Bot. Gaz., xv. 243.

Frequent in very wet places, from the sea level to mountain summits. Appearing in a number of forms distinguished merely by color, black (*f. nigricans,* Brid.), green (*f. virescens,* Warnst.), yellow (*f. flava,* Warnst.), and red-brown (*f. rufescens,* Warnst.). Green Mt.; Browns Mt. (E. Faxon); — Sea Wall; Sunken Heath; Aunt Bettys Pond (Faxon & Rand); — Breakneck Ponds; Pemetic Mt. (Rand).

S. subsecundum, Nees. L. & J., Mosses N. A., 19.

Meadow on Denning Brook; Town Hill; Sea Wall Swamp; Southwest Valley road (E. Faxon); — Bass Harbor Marsh (Rand).

Var. **macrophyllum,** Roell.

Stem leaves lingulate, evenly bordered, 1.14–1.45 mm. long, nearly two thirds as broad, the upper part porose and fibrillose; hyaline cells often partitioned ; branch leaves ovate, 1.71–1.90 mm. long, little more than half as broad, inner pores in the cell angles, mostly near the margins. Southwest Harbor ; Upper Breakneck Pond (Rand); — Little Cranberry Isle (Redfield).

Var. **mesophyllum,** Warnst.

Stem leaves lingulate, evenly bordered, 1–1.40 mm. long, two thirds to four fifths as broad, the upper part porose and fibrillose; hyaline cells much partitioned; branch leaves ovate, about as long and broad as the stem leaves, inner surface with feebly ringed pores towards the apex. Breakneck Ponds; Beech Mt. Notch; woods, Norwood Cove; Deer Brook (Rand).

Var. **microphyllum,** Roell.

Stem leaves small, somewhat triangular, border widened to the base, about 0.57 mm. long, and the same in greatest breadth, usually without pores or fibrils; hyaline cells not partitioned; branch leaves lance-ovate, about 0.80 mm. long, and little more than half as broad, loosely somewhat secund, inner surface without pores. Southwest Harbor (Rand).

S. rufescens, Bryol. Germ. Bot. Gaz., xv. 246. *S. subsecundum,* Nees, var. *contortum,* Schimp. L. &. J., Mosses N. A., 19.

In water. Upper Breakneck Pond (Faxon & Rand); — border of Somes Pond (E. Faxon).

§ 7. CYMBIFOLIA.

S. imbricatum (Hornsch.), Russ. Bot. Gaz., xv. 249. *S. Austini,* Sulliv. L. & J., Mosses N. A., 21. Var. **sublæve,** Warnst. Bot. Gaz., xv. 250.

Woods, The Barcelona meadow (Rand); — Gilmore Meadow ; Little Harbor Brook (Redfield).

Var. **cristatum,** Warnst. Bot. Gaz., xv. 250.

Sea Wall Swamp (E. Faxon); — Breakneck Ponds ; bogs, southwest of Sea Wall, and north of Beech Hill (Rand).

Var. **affine** (Ren. & Card.), Warnst. Bot. Gaz., xv. 250.

Swamp on Meadow Brook, Somesville; Clark Cove; The Barcelona meadow (E. Faxon); — bog north of Beech Hill; Sutton Island; Beech Mt. Notch; Deer Brook (Rand).

S. cymbifolium, Ehrh. L. & J., Mosses N. A., 21.

Canada Valley; bogs, southwest of Sea Wall; Southwest Harbor; Sutton Island, and elsewhere (Rand).

Var. **glaucescens,** Warnst. Bot. Gaz., xv. 251.

Cold Brook (Rand).

S. papillosum, Lindb. L. & J., Mosses N. A., 21. *S. cymbifolium,* Ehrh., var. *papillosum* (Lindb.), Schimp. Bot. Gaz., xv. 251. Var. **normale,** Warnst.

Hadlock Upper Pond ; Great Cranberry Isle (Rand); — Sunken Heath (E. Faxon).

S. medium, Limpr. Bot. Gaz., xv. 252.

Common. Beech Hill road (Faxon & Rand); — bogs southwest of Sea Wall, and north of Beech Hill (Rand).

Var. **purpurascens**, Russ. *S. medium*, var. *læve, f. purpurascens* (Russ.), Warnst. Bot. Gaz., xv. 253.

Somes Pond (E. Faxon); — Aunt Bettys Pond; bogs southwest of Sea Wall; Red Rocks, Great Cranberry Isle (Rand); — Little Cranberry Isle (Redfield).

Var. **pallescens**, Warnst.

Color very pale yellowish-green. Bog on Prettymarsh road, west of Somesville (Rand).

ORDER II. ANDREÆACEÆ. SCHIZOCARPOUS MOSSES.

ANDREÆA, Ehrh.

A. petrophila, Ebrh.

Frequent on wet rocks on the mountains. Sargent Mt. (Faxon & Rand); — Green Mt. (D. C. Eaton); — Beech Mt. (Rand); — Robinson Mt. (E. Faxon).

A. crassinervis, Bruch.

On wet rocks, summit of Sargent Mt. (E. Faxon); — Beech Mt.; Beech Cliff (Faxon & Rand).

ORDER III. BRYACEÆ. TRUE MOSSES.

List prepared by Edward L. Rand, under the supervision of Elizabeth G. Britton. Plants collected mainly by Walter L. Burrage, John H. Redfield, Edward L. Rand, Edwin Faxon, and Theodore G. White, determinations by Elizabeth G. Britton and Dr. Charles R. Barnes.

In view of the great importance of following the arrangement and nomenclature adopted in some standard work of ready reference, and in view also of the great disagreement on these very matters of arrangement and nomenclature among the authorities, it has seemed wise to follow Lesquereux and James's "Mosses of North America" in the preparation of this list. Synonyms, however, have been given where they seemed of real value; some necessary descriptions have been added; and a few corrections made where they were of real, not merely of verbal importance.

SECTION I. ACROCARPI.

Tribe **WEISIEÆ**.

CYNODONTIUM, Schimp.

C. polycarpum (Ehrh.), Schimp. *Oncophorus polycarpus* (Ehrh.), Brid.

On decayed wood. Wood road to Great Pond, Southwest Harbor (Rand).

C. virens (Swz.), Schimp., var. **Wahlenbergii** (Brid.), Schimp. *Oncophorus Wahlenbergii*, Brid.

On decayed wood. Upper Breakneck Pond (Rand).

TREMATODON, Mx.

T. ambiguum (Hedw.), Hornsch.

On the ground. Green Mt. (D. C. Eaton); — moist woods, north of Long Pond (Theodore G. White).

DICRANELLA, Schimp.

D. squarrosa (Starke), Schimp.

In dense mats on sandy shore, Jordan Pond; submersed (Rand). As the specimens found are all sterile, some doubt has arisen as to their identity. They are almost identical with specimens collected by E. Faxon at Ammonoosuc Lake, Crawfords, N. H., June 14, 1883, and very similar to specimens collected by Oakes at the White Mts., N. H. (Sulliv. & Lesq., Musci Bor. Am., No. 245). The latter specimens were distributed as "*Meesia longiseta*, Hedw. var. ?" Faxon's specimens, however, are *Dicranella squarrosa*, and so with very little doubt are the Mt. Desert specimens. The true identity of Oakes's specimens is still in doubt. Especial thanks are due Dr. C. R. Barnes and Mrs. E. G. Britton for the solution of these puzzling questions of identification.

D. heteromalla (L.), Schimp.

On the ground ; common (Rand). Also Great Cranberry Isle (Theodore G. White).

DICRANUM, Hedw.

D. Blyttii, Bruch & Schimp. *D. Schisti* (Gunn.), Lindb. Falls, Sargent Mt. Gorge (Theodore G. White).

D. montanum, Hedw.

On decaying trees. Woods, Hadlock Upper Pond (Walter L. Burrage); — near Ripples Pond; Beech Mt.; Oak Hill (Rand).

D. viride (Sull. & Lesq.), Lindb.

On decaying trees; sterile. Near Aunt Bettys Pond; Cold Brook; Seal Harbor (Rand); — on ledges, Seal Harbor (Theodore G. White).

D. flagellare, Hedw.

On decaying tree trunks; common (W. L. Burrage, Rand, E. Faxon, T. G. White).

D. fulvum, Hook.

On granitic rocks, West Branch of Hadlock Brook; Southwest Valley road (Rand). On ground, woods, head of The Barcelona meadow (Rand); — also Seal Harbor (Theodore G. White).

D. longifolium, Ehrh.

Dry woods, behind schoolhouse, Seal Harbor (Theodore G. White).

D. fuscescens, Turn.

On decaying tree trunks; common (Rand, Theodore G. White).

Var. **longirostre,** Schimp.

Woods on Sargent Mt. (Walter L. Burrage).

D. congestum, Brid. *D. fuscescens,* Lesq. & J., in part.

On decayed wood. Northern foot of Beech Mt.; Upper Breakneck Pond; Jordan Pond trail, Northeast Harbor (Rand).

D. scoparium (L.), Hedw.

On ground, rocks, etc.; common (Walter L. Burrage, Rand, Theodore G. White); — also Great Cranberry Isle (Theodore G. White).

Var. squarrosum, Lesq. & J.

Seal Harbor (Theodore G. White).

Var. paludosum, Bruch & Schimp.

Seal Harbor (Theodore G. White).

Var. pallidum, C. Mueller.

Jordan Pond trail, Seal Harbor (Theodore G. White).

Var. rupestre, Austin.

Leaves short, curled and twisted when dry; plants small and usually sterile. Musci App., No. 90 (1870). S. & L., Musci Bor. Am., ed. 2, No. 76 (1865). On rocks in woods. Breakneck Ponds (Annie S. Downs); — Seal Harbor (Theodore G. White).

Var. recurvatum (Schultz), Brid.

Plants tall and slender, usually sterile, bright yellow; leaves uncinate, recurved, narrow and plumose at tip; distant, not crowded on the stems. Prettymarsh Harbor (Theodore G. White).

D. majus, Smith.

Damp ground. Cold Brook; Great Cranberry Isle (Rand); — Sutton Island (Theodore G. White).

D. palustre, Bruch & Schimp. *D. Bonjeani*, DeNot.

Moist ground. Sargent Mt. (Faxon & Rand); — Western Mt.; Beech Mt. Notch; woods, Norwood Cove; wood road to Great Pond, Southwest Harbor; Seal Harbor (Rand); — Prettymarsh; The Cleft; Seal Harbor (Theodore G. White).

D. Schraderi, Web. & Mohr. *D. Bergeri*, Bland.

Moist ground in woods and on the mountains. Browns Mt. (Faxon & Rand); — Dog Mt. (E. Faxon); — Norwood Cove; Intervale Brook; Beech Cliff; near Aunt Bettys Pond (Rand).

D. spurium, Hedw.

Dry ground. Browns Mt. (Redfield, Faxon & Rand); — Browns Mt. Notch; Seal Harbor; West Branch of Hadlock Brook (Rand); — Barr Hill (Theodore G. White).

D. brachycaulon, Kindb.

Allied to *D. spurium*, but differing in the short stem only
1 cm. high, the leaves smaller and shorter, oblong-ovate, acute,
not acuminate, entire, not papillose at back, costa elevate, per-
current and smooth, alar cells brown, capsule small, pedicel
1 cm. long. Peculiar in its short leaves and its elevate costa.
Kindberg in Macoun, Cat. Canadian Plants, part vi. 34.
Not uncommon in dry places on the hills and mountains.
High Head (Faxon & Rand); — Dog Mt.; Robinson Mt. (E.
Faxon); — Sargent Mt.; Jordan Mt. (Rand).

D. Drummondi, C. Mueller.

On ground, mostly in woods; frequent. Browns Mt. (E.
Faxon); — Western Mt.; Norwood Cove; Upper Breakneck
Pond (Rand); — between Hadlock farm and Frenchman Camp
(Redfield); — Seal Harbor; High Head (Theodore G. White).

D. undulatum, Ehrh.

On ground in woods; very common (A. B. Eaton, Burrage,
Rand, Faxon, Redfield, White). An unusual form, with five
pedicels in a cluster, Beech Cliff (Theodore G. White).

FISSIDENS, Hedw.

F. adiantoides (L.), Hedw.

On wet rocks. Western Mt.; Southwest Valley road (Rand).

LEUCOBRYUM, Hampe.

L. vulgare, Hampe. *L. glaucum* (L.), Schimp.

On ground, woods and hills; common (Rand, Redfield, White).

L. minus, Sulliv. *L. albidum* (Brid.), Lindb.

On ground. Great Cranberry Isle; Seal Harbor (Theodore
G. White).

CERATODON, Brid.

C. purpureus (L.), Brid.

On rocks and ground ; common (Burrage, Faxon, Rand,
White). Also Great Cranberry Isle (Theodore G. White).

C. conicus (Hampe), Lindb.

"Differs from *C. purpureus* in the long excurrent costa, the capsule erect and symmetric, faintly sulcate, the lid shorter, the teeth pale, red at base, with fewer articulations." Kindberg in Macoun, Cat. Canadian Plants, part vi. 39.* On rocks. Browns Mt. (Walter L. Burrage, Faxon & Rand).

Tribe POTTIEÆ.

LEPTOTRICHUM, Hampe.   (*Ditrichum*, Timm.)

L. pallidum (Schreb.), Hampe.   *D. pallidum* (Schreb.), Hampe.

A depauperate form on ground, Long Pond, Eden (Rand).

BARBULA, Hedw.

B. tortuosa (L.), Web. & Mohr.

On ground.   West Branch of Hadlock Brook (Rand).

Tribe GRIMMIEÆ.

GRIMMIA, Ehrh.

G. conferta, Funck.

On rocks.   Seal Harbor (Theodore G. White).

G. apocarpa (L.), Hedw., var. **gracilis** (Schleich.), Web. & Mohr.

On rocks.   East Point, Seal Harbor (Theodore G. White).

RACOMITRIUM, Brid.

R. aciculare (L.), Brid.

Wet rocks, in brooks, etc.   Little Harbor Brook (Redfield, Rand); — Intervale Brook (Rand).

R. heterostichum (Hedw.), Brid.

On rocks.   Beech Cliff (Rand); — Dog Mt. (E. Faxon).

* See also Braithw., British Moss Flora, 175.

R. fasciculare (Schrad.), Brid.

On rocks. Browns Mt.; Sargent Mt. (Faxon & Rand); — The Cleft (Theodore G. White).

R. microcarpum, Brid.

On rocks. Browns Mt.; Beech Mt.; north of Beech Hill (Rand); — Robinson Mt. (E. Faxon).

R. lanuginosum, Brid. *R. hypnoides* (L.), Lindb.

On rocks, mountain summits. Sargent Mt. (E. Faxon); — Green Mt. (D. C. Eaton); — Pemetic Mt. (Rand).

HEDWIGIA, Ehrh.

H. ciliata, Ehrh.

On rocks; common. Sargent Mt., and elsewhere (Faxon & Rand); — Beech Hill; Somesville, and elsewhere (Rand); — Seal Harbor (Theodore G. White).

Var. **viridis**, Schimp.

Browns Mt. (Redfield); — Seal Harbor (Theodore G. White).

Tribe ORTHOTRICHEÆ.

AMPHORIDIUM, Schimp.

A. Lapponicum (Hedw.), Schimp.

On wet rocks. Canada Cliff (E. Faxon).

A. Mougeotii (Benth.), Schimp.

On rocks. Green Mt. Gorge (D. C. Eaton).

ULOTA, Mohr. (*Weissia*, Ehrh.*)

U. Ludwigii, Brid. *Weissia coarctata* (Beauv.), Lindb.

On trees; common (Burrage, Faxon, Rand, Redfield, White).

U. crispa, Brid. *Weissia ulophylla*, Ehrh.

On trees; common (Burrage, Rand, Redfield). On rocks. Balance Rock, Seal Harbor; north of Beech Hill (Rand).

* See Elizabeth G. Britton, N. A. Species of Weissia, Bull. Torr. Bot Club, xxi. 65.

U. crispula, Brid.   *Weissia crispula,* Lindb.

On trees.   Sargent Mt. (Faxon & Rand).

U. phyllantha, Brid.   *Weissia phyllantha* (Brid.), Lindb.

On trees.   Seal Harbor (Theodore G. White).

U. Hutchinsiæ (Smith), Hammar.   *Weissia Americana* (Beauv.), Lindb.

On rocks; common (Redfield, Rand, White).

ORTHOTRICHUM, Hedw.

O. fallax, Schimp.   *O. Schimperi,* Hammar.

On willow trees.   Somesville (Rand).

Tribe TETRAPHIDEÆ.

TETRAPHIS, Hedw.   (*Georgia,* Ehrh.)

T. pellucida (L.), Hedw.   *Georgia pellucida* (L.), Ehrh.

On decaying wood; common (Burrage, Rand, White).

Tribe SPLACHNEÆ.

SPLACHNUM, L.

S. ampullaceum, L.

On cow dung.   Bog hole on wood road to Aunt Bettys Pond, Youngs District (Rand, Faxon).   On rotten wood.   Woods north of Long Pond (Theodore G. White).

Tribe PHYSCOMITRIEÆ.

FUNARIA, Schreb.

F. hygrometrica (L.), Sibth.

On ground; common, especially on burnt soil (Rand, White).

Tribe BARTRAMIEÆ.

BARTRAMIA, Hedw.

B. pomiformis (L.), Hedw.

Rocks and shady banks. Sargent Mt. Gorge (Walter L. Burrage); — Beech Mt. Notch; High Head; Western Mt. (Rand); — Triad Pass (Theodore G. White).

Var. **crispa** (Swz.), Schimp.

Browns Mt. Notch (Rand) ; — between Hadlock farm and Frenchman Camp (Redfield).

PHILONOTIS, Brid.

P. Muhlenbergii, Brid.

Wet places. Sargent Mt.; Browns Mt. (Faxon & Rand); — Beech Hill (Rand).

P. fontana (L.), Brid.

Wet places and moist rocks; common (Rand, Faxon, White).

Tribe BRYEÆ.

LEPTOBRYUM, Schimp.

L. pyriforme (L.), Wils.

Ox Hill, Seal Harbor (Theodore G. White).

WEBERA, Hedw. (*Pohlia*, Hedw.)

W. nutans (Schreb.), Hedw. *Pohlia nutans* (Schreb.), Lindb.

On ground, and rock hollows on mountains; common (Faxon, Rand, White).

BRYUM, L.

B. bimum, Schreb.

Moist places. Somesville (Redfield, Rand); — shore of North-west Arm, Great Pond (Rand).

B. cæspiticium, L.

On ground. Long Pond, Eden; High Head (Rand); — Seal Harbor (Theodore G. White).

B. capillare, L.

In rich soil. Triad Pass; banks of Somes Stream (Rand).

B. pseudotriquetrum (Hedw.), Schwaegr. *B. ventricosum,* Dicks.

Shores. Ripples Pond; Northwest Arm, Great Pond (Rand).

MNIUM, L. (*Astrophyllum,* Neck.)

M. cuspidatum, Hedw. *Astrophyllum sylvaticum,* Lindb.

On ground, shady places; common (Rand, White).

M. rostratum, Schwaegr. *Astrophyllum rostratum* (Schrad.), Lindb.

Seal Harbor (Theodore G. White).

M. affine, Bland. *Astrophyllum cuspidatum* (L.), Lindb.

In moist ground, shady places by watercourses, etc; common (Burrage, Rand).

M. hornum, L. *Astrophyllum hornum* (L.), Lindb.

Moist, shady woods. Seal Harbor (Redfield); — Otter Cliffs; High Head; Norwood Cove; head of The Barcelona meadow; Somesville (Rand); — Barr Hill; Great Cranberry Isle; Sutton Island (Theodore G. White).

M. cinclidioides (Blytt), Hueben. *Astrophyllum cinclidioides* (Blytt), Lindb.

Woods, Beech Hill road, Southwest Harbor (Rand).

M. punctatum (L.), Hedw. *Astrophyllum punctatum* (L.), Lindb.

Moist woods and wet places ; common (Rand, Redfield, White). A small form, Browns Mt. Notch (Rand).

AULACOMNIUM, Schwaegr. (*Sphærocephalus*, Neck.)

A. palustre (L.), Schwaegr. *Sphærocephalus palustris* (L.), Lindb.

Marshy places; common (Burrage, Rand, White).

Var. **polycephalum**, Bruch & Schimp.

Frequent. Near Northwest Cove (E. Faxon); — Beech Hill; near Ripples Pond, and elsewhere (Rand); — near Long Pond (Theodore G. White).

Tribe POLYTRICHEÆ.

ATRICHUM, Beauv. (*Catharinea*, Ehrh.)

A. undulatum (L.), Beauv. *Catharinea undulata* (L.), Web. & Mohr.

In damp ground. Beech Mt. Notch; brookside, Southwest Valley road ; Stanley Brook ; Somesville (Rand) ; — Great Cranberry Isle (Theodore G. White).

A. angustatum, Bruch & Schimp. *Catharinea angustata*, Brid.

On ground. Beech Mt. Notch; High Head; Seal Harbor; Southwest Valley road (Rand).

POGONATUM, Beauv.

P. brevicaule, Beauv. *P. tenue* (Menz.), E. G. Britton.

On clay banks. Wood road to Great Pond, Southwest Harbor; Beech Cliff road, Somesville (Rand); — Southwest Harbor (M. L. Fernald).

P. alpinum (L.), Roehl.

On rocks. Browns Mt. (Walter L. Burrage); — Ovens ; Browns Mt. Notch (Rand); — slope of Green Mt. (D. C. Eaton); — Triad Pass (Theodore G. White).

POLYTRICHUM, L.

P. Ohioense, Ren. & Card.   *P. formosum*, Lesq. & J., Mosses N. A., 264 in part.

Stem erect, simple or bipartite, 3–6 cm. long, a little tomentose below; leaves spreading when moist, erect-flexuous when dry, from a sheathing base, linear-acuminate, cuspidate, serrate; lamellæ about 50, each in section of a row of 5–7 cells, the marginal one much larger, transversely dilated, about twice broader than high, very slightly convex, often almost plane; perichætial leaves longer, with a longer hyaline base. Pedicel 4–8 cm. long, reddish below, pale above ; capsule erect, finally horizontal, tetragonal or pentagonal, rarely hexagonal, acute-angled, rather narrowed toward the base, with a very small or indistinct hypophysis; length 5–7 mm., diameter 2–2½ mm.; lid conic-acuminate, red at margin. Distinguished from *P. formosum* by the form of the capsule, more or less narrowed toward the base, and with an indistinct hypophysis; and further chiefly by the form of the marginal cells of the lamellæ. Ren. & Card., Rev. Bryol. (1885), 11. Bot. Gaz., xiii. 199. Macoun, Cat. Canadian Plants, part vi. 153. On ground. Woods, Hadlock Valley (Redfield).

P. piliferum, Schreb.

Dry grounds, and in rock hollows. Sargent Mt.; Browns Mt. (Walter L. Burrage); — Flying Mt. (R. & R.); — Newport Mt. (Theodore G. White).

P. juniperinum, Willd.

Dry, open ground. Dog Mt.; Browns Mt.; Asticou Hill (Walter L. Burrage); — Otter Creek quarries (Theodore G. White).

P. alpinum, L.   *P. juniperinum*, Willd., var. *alpinum* (L.), Schimp.

Barr Hill, and elsewhere (Theodore G. White).

P. strictum, Banks.

Dry ground. Browns Mt. (Rand); — Southwest Harbor (A. B. Eaton); — Salisbury Cove (Walter L. Burrage); — Long Pond (Theodore G. White).

P. commune, L.

Dry or moist ground, woods and open places; very common (Burrage, Rand, White).

Tribe BUXBAUMIEÆ.

DIPHYSCIUM, Mohr. (*Webera*, Ehrh.)

D. foliosum (Web.), Mohr. *Webera sessilis* (Schmid.), Lindb.

On clay bank, shore south of Aunt Mollys Beach (Rand).

SECTION II. PLEUROCARPI.

Tribe FONTINALEÆ.

FONTINALIS, L.

F. antipyretica, L., var. gigantea, Sulliv.

Frequent on stones in brooks. Cold Brook; brook, Clark Valley, and elsewhere (Rand); — Bubble Pond (Redfield).

F. Dalecarlica, Bruch & Schimp.

On stones in brooks; common (Burrage, Rand). Also in still water on the shore of Great Pond (Rand).

F. Novæ-Angliæ, Sulliv.

Rivulet flowing into Denning Pond (E. Faxon); — Somes Stream (Rand); — Hunters Brook (Theodore G. White).

F. Lescurii, Sulliv.

Doctors Brook (Redfield).

F. Sullivantii, Lindb.

In swift and still water. Long Pond, Eden (E. Faxon); — runlet near head of Denning Pond (Rand).

DICHELYMA, Myrin.

D. pallescens, Bruch & Schimp.

On twigs in wet hole in woods, head of The Barcelona meadow (Rand).

Tribe NECKEREÆ.

NECKERA, Hedw.

N. pennata (L.), Hedw.

On trees; frequent. Seal Harbor (Redfield); — Canada Valley; Upper Breakneck Pond, etc. (Rand); — slopes of Green Mt. (D. C. Eaton); — Southwest Harbor (M. L. Fernald); — Barr Hill (Theodore G. White).

Tribe LEUCODONTEÆ.

LEUCODON, Schwaegr.

L. sciuroides (L.), Schwaegr.

On willow trees, Somesville (Rand).

Tribe LESKEEÆ.

MYURELLA, Bruch & Schimp.

M. julacea (Vill.), Bruch & Schimp.

On ground. Southwest Harbor (Rand).

LESKEA, Hedw.

L. tristis, Cesati.

On trees. Deer Brook, near Jordan Pond (Rand).

Tribe ORTHOTHECIEÆ.

PLATYGYRIUM, Bruch & Schimp.   (*Entodon*, C. Mueller.)

P. repens, Bruch & Schimp.   *E. palatinus* (Neck.), Lindb.

On willow trees, Somesville (Rand).

PYLAISIA, Bruch & Schimp.

P. polyantha (Schreb.), Bruch & Schimp.

On old plank, Southwest Harbor (Rand).

P. velutina, Bruch & Schimp.

On trees. Canada Valley; West Branch of Hadlock Brook; Deer Brook; Intervale Brook (Rand).

CLIMACIUM, Web. & Mohr.

C. dendroides (L.), Web. & Mohr.

On ground. Woods, head of The Barcelona meadow; Clark Valley (Rand).

C. Americanum, Brid.

On ground. Moist thicket, Somes Stream (Rand).

Tribe HYPNEÆ.

HYPNUM, L.

Subgenus THUIDIUM.

H. recognitum, Hedw.

On ground, rocks, etc.; common (Walter L. Burrage, Rand).

H. delicatulum, L.

On ground, rocks, etc. Western Mt.; Cold Brook (Rand); — Triad Pass (Theodore G. White).

Subgenus BRACHYTHECIUM.

H. lætum, Brid.

On trees. West Branch of Hadlock Brook (Rand).

H. salebrosum, Hoffm. *H. plumosum,* Huds.

On trees. Roadside near Ripples Pond; near Seal Harbor reservoir; Clark Valley (Rand).

Var. **palustre,** Lesq. & J.

Wet ground. Woods, head of The Barcelona meadow; junction of Prettymarsh and Seal Cove roads (Rand).

H. velutinum, L.

On ground. Dry woods, Seal Harbor (Theodore G. White).

H. rutabulum, L.

On rocks or stumps. North of Beech Mt.; Northwest Arm woods (Rand); — on Jordan Stream (Theodore G. White).

H. campestre, Bruch.

On ground, rocks, and logs. Near head of Denning Pond; Southwest Harbor; Intervale Brook; woods, Upper Breakneck Pond; Western Mt. (Rand).

H. Novæ-Angliæ, Sulliv. & Lesq.

On ground. Asticou Hill; Baker Island (Theodore G. White); — Seal Harbor; Cold Brook (Rand).

H. plumosum, Swz. *H. pseudoplumosum*, Brid.

On rocks. Road to Beech Hill, Somesville (Rand); — Newport Mt. (Theodore G. White).

Subgenus EURHYNCHIUM.

H. strigosum, Hoffm.

Bog hole, Prettymarsh road west of Ripples Pond (E. Faxon).

H. Sullivantii, Spruce.

On ground. Woods at Salisbury Cove (Walter L. Burrage).

Subgenus RAPHIDOSTEGIUM.

H. recurvans (Mx.), Schwaegr.

On tree roots and old logs. Foot of Western Mt.; Salisbury Cove (Walter L. Burrage); — woods, Norwood Cove ; Upper Breakneck Pond (Rand).

H. cylindricarpum, C. Mueller.

On trees. Browns Mt. (R. & R.).

H. Jamesii, Lesq. & J.

On ground. Triad Pass (Theodore G. White).

Subgenus RHYNCHOSTEGIUM.

H. deplanatum, Schimp.

On decaying wood. Jordan Pond trail from Northeast Harbor (Rand).

H. serrulatum, Hedw.

On wet rocks. Western Mt. On wet ground. Near Ripples Pond (Rand).

H. rusciforme, Weis.

Wet rocks. East Branch of Hadlock Brook (Walter L. Burrage); — Browns Mt. (Redfield); — Sargent Mt.; West Branch of Hadlock Brook (Rand).

Subgenus PLAGIOTHECIUM.

H. micans, Swz.

On decaying wood. Seal Harbor (Theodore G. White).

H. turfaceum, Lindb.

On ground and old logs in woods; common (Redfield, Rand, White).

H. elegans, Hook.

Crevices of rocks. Dog Mt. (Rand); — Seal Harbor (Theodore G. White).

H. denticulatum, L.

On decaying tree trunks. Browns Mt. Notch; Norwood Cove (Rand); — Browns Mt. (Redfield); — woods north of Long Pond (Theodore G. White). On rocks. Northern foot of Beech Mt. (Rand).

H. Muhlenbeckii, Spruce. *H. striatellum,* Brid.

On rocks; frequent (Rand).

Subgenus AMBLYSTEGIUM.

H. serpens, L.

On decaying wood. Southwest Harbor (Rand).

H. orthocladon, Beauv.

On old trees. Woods, Intervale Brook (Rand).

H. Lescurii, Sulliv.

On wet rocks. Intervale Brook, near bridge (Rand).

H. riparium, L.

On decaying wood, Stanley Brook (Rand); — immersed among *Fontinalis Dalecarlica,* shore of Northwest Arm, Great Pond (Rand).

Var. flaccidum, Lesq. & J.

Deep woods, Little Cranberry Isle (Redfield).

SUBGENUS CAMPYLIUM.

H. hispidulum, Brid.

On wet rocks.   Western Mt. (Rand).

H. chrysophyllum, Brid.

On tree roots in moist ground.   Long Pond (Theodore G. White); — meadow on Sunken Heath Brook (Rand).

Var. rupestre, Aust.

Plants large, in dense cushions; stems usually long, branching; generally sterile.   Austin, Musci App., No. 396 (1870).   On rocks in brooks.   Sargent Mt. Gorge (Walter L. Burrage).

Var. cæspitosum, Aust.

Plants slender, densely cæspitose.   Austin, Musci App., No. 395 (1870).   About tree roots.   Seal Harbor (Theodore G. White).

H. polygamum, Wils.

In wet ground.   Somesville (Rand).

SUBGENUS HARPIDIUM.

H. aduncum, Hedw.

In wet ground.   Sea Wall; Beech Cliff; Somes Pond (Rand).

H. uncinatum, Hedw.   *H. aduncum,* L.

On ground, rocks, and decaying wood; very common.   (Burrage, Rand, Redfield.)

Var. gracilescens, Bruch & Schimp.

On ground in woods, Great Cranberry Isle (Rand).

H. fluitans, L.

In ditches, bog holes, and wet places. Near Spruce Point, Eden (E. Faxon); — Sea Wall (Redfield); — Ripples Pond ; Pemetic Mt.; Jordan Mt.; Browns Mt., a very slender form (Rand).

SUBGENUS CTENIUM.

H. Crista-castrensis, L.

On ground and old logs; frequent. Upper Hadlock Pond (Walter L. Burrage); — Sargent Mt. (Faxon & Rand); — High Head (H. S. Rand); — Southwest Valley road, etc. (Rand); — between Hadlock farm and Frenchman Camp (Redfield); — Great Head (Theodore G. White).

SUBGENUS HYPNUM, PROPER.

H. reptile, Mx.

On ground or on trees. South end of Great Pond; Cold Brook; Southwest Valley road; Western Mt.; Beech Mt.; Northwest Arm woods; Intervale Brook (Rand).

H. fertile, Sendt.

On moist rocks, old logs, etc. Browns Mt. Notch; Intervale Brook (Rand); — woods, north of Long Pond (Theodore G. White).

H. imponens, Hedw.

On ground, old logs, etc. Canada Valley; Norwood Cove; Northwest Arm woods; Intervale Brook (Rand); — The Cleft (Theodore G. White).

H. cupressiforme, L.

On rocks, old trees, etc.; common and variable. Hadlock Upper Pond (Walter L. Burrage); — south end of Great Pond, and elsewhere (Rand); — Seal Harbor; Sutton Island (Theodore G. White).

Var. filiforme, Brid.

On Jordan Stream (Theodore G. White).

Var. ericetorum, Bruch & Schimp.

Northwest Arm woods (Rand).

Var. resupinatum (Wils.), Schimp.

Cæspitose, pale green ; leaves falcate or curved, yellow at angles; capsule suberect or incurved, lid rostrate.   On Jordan Stream (Theodore G. White); — woods, Intervale Brook (Rand).

H. curvifolium, Hedw.

Wet rocks and old logs.   Waterfall, East Branch of Hadlock Brook (Theodore G. White) ; — Somes Pond; Intervale Brook (Rand).

H. pratense, Koch.

In wet ground, Seal Cove road, Southwest Harbor (Rand).

H. Haldanianum, Grev.

Wet clayey ground and on tree trunks.   Ripples Pond; between Bass Harbor and Southwest Harbor (Rand); — Prettymarsh (Theodore G. White).

SUBGENUS LIMNOBIUM.

H. palustre, Huds.

On rocks.   Woods on Intervale Brook (Rand).

H. molle, Dicks.   H. dilatatum, Wils.

On wet rocks.   Brook, Western Mt. (Walter L. Burrage); — Browns Mt. Notch (Rand).

H. eugyrium, Schimp.

Woods, north of Long Pond (Theodore G. White); — Intervale Brook (Rand).

H. ochraceum, Turn.

In cold mountain brooks.   North of Sargent Mt.; near head of Denning Pond (Rand); — at waterfall, East Branch of Hadlock Brook (Theodore G. White).   A very slender form, head waters of Gilmore Brook (E. Faxon).

Subgenus CALLIERGON.

H. cordifolium, Hedw.

Common in swamps and wet boggy ground.   (E. Faxon, Rand, White.)

H. Schreberi, Willd.   *H. parietinum*, L.

On ground; common.   (Burrage, E. Faxon, Rand, Redfield, White.)

H. stramineum, Dicks.

Among sphagnum in bogs.   Little Cranberry Isle (Redfield); — Breakneck Ponds; Great Cranberry Isle (Rand).

Subgenus PLEUROZIUM.

H. splendens, Hedw.   *H. proliferum*, L.

Moist rocks and on ground in woods; common.   (Burrage, Redfield, Rand, White.)

H. brevirostre, Ehrh.

Gorge, West Branch of Hadlock Brook (Rand).

Subgenus HYLOCOMIUM.

H. squarrosum, L.

On ground.   Cold Brook (Rand).

H. triquetrum, L.

Woods, on ground; common.   (Burrage, Rand.)

Division II.   HEPATICÆ; LIVERWORTS.

List prepared by Edward L. Rand under the supervision of Dr. L. M. Underwood, by whom many annotations and some necessary descriptions have been furnished.   Plants collected by Edward L. Rand and others; determined by Dr. Underwood. The arrangement followed is that given in the sixth edition of Gray's Manual.

JUNGERMANNIACEÆ. Scale Mosses.

FRULLANIA, Raddi.

F. Eboracensis, Lehm.

Common on trees; sometimes on rocks.

F. Asagrayana, Mont.

On spruce trees, Beech Mt. (Rand); — on rocks, near Little Harbor (Redfield) ; — among moss on old log, Great Pond (Rand).

JUBULA, Dumort.

J. Hutchinsiæ (Hook.), Dumort., var. **Sullivantii,** Spruce.

Wet rocks in Intervale Brook (Rand).

PORELLA, L.

P. platyphylla (L.), Lindb. *Madotheca platyphylla,* Dumort.

Frequent, usually on trees. Seal Harbor (Redfield); — Deer Brook; Hadlock Brook; Breakneck road (Rand).

PTILIDIUM, Nees.

P. ciliare (L.), Nees.

Very common on ground, rotten trees, etc. Variable. An unusually small form on old log, path to Beech Cliff (Rand). A large form, among sphagnum in bog hole on The Heath, Great Cranberry Isle, appears to be *P. pulcherrimum* (Web.), Nees.

TRICHOCOLEA, Dumort.

T. tomentella (Ehrh.), Dumort.

In moss, Cold Brook (Rand).

BAZZANIA, S. F. Gray.

B. trilobata (L.), S. F. Gray. *Mastigobryum trilobatum,* Nees.

Very common on ground in damp woods. A very delicate form on wet rocks, northern end of Beech Mt. (Rand).

LEPIDOZIA, Dumort.

L. reptans (L.), Dumort.

On wet rocks among moss, northern end of Beech Mt.; on ground, Norwood Cove (Rand).

L. setacea (Web.), Mitt.

In sphagnum, border of Aunt Bettys Pond (Faxon & Rand). This seems to be a floating form.

BLEPHAROSTOMA, Dumort.

B. trichophyllum (L.), Dumort.

On clay banks, Stanley Brook, Seal Harbor (Rand).

CEPHALOZIA, Dumort.

C. Virginiana, Spruce.

In wet ground. Southern end of Great Pond (Rand).

C. multiflora, Spruce.

In wet ground. Sunken Heath; Sunken Heath Brook; Beech Mt. (Rand).

C. bicuspidata (L.), Dumort.

On the ground. Jordan Pond trail from Northeast Harbor; Norwood Cove (Rand).

C. curvifolia (Dicks.), Dumort.

On rotten logs. Northwest Arm woods (Rand).

C. fluitans (Nees), Spruce.

Very abundant in shallow pools and among sphagnum, Sunken Heath (Faxon & Rand); — Sunken Heath Brook (Rand).

C. divaricata (Sm.), Dumort.

On rocks with Andreæa, Sargent Mt. (Rand).

KANTIA, S. F. Gray.

K. Trichomanis (L.), S. F. Gray.

On ground. Wood road to Western Mt.; Jordan Pond path, Seal Harbor (Rand).

SCAPANIA, Dumort.

S. undulata (L.), Dumort.

On stones in brooks, usually submersed; frequent.

S. irrigua (Nees), Dumort.

On rocks in water. Jordan Pond; Stanley Brook (Rand).

S. nemorosa (L.), Dumort.

On wet rocks and damp ground; very common.

DIPLOPHYLLUM, Dumort.

D. albicans (L.), Dumort.

This species is distinguished from *D. taxifolium* "by the presence in the two lobes of a pseudo-nerve, which is often colorless, and consists of a series of from 4 to 6 elongated cells. A cross-section of the leaf shows the cells to be of equal diameter as the others, only with the outer walls thickened considerably." Pearson, Canadian Hepaticæ, 15. On rocks. Browns Mt. Notch (Rand).

D. taxifolium (Wahl.), Dumort. *D. albicans*, Dumort., var. *taxifolium*, Nees. Gray, Man., 6th ed., 715.

On rocks. West Branch of Hadlock Brook (Rand).

D. Dicksoni (Hook.), Dumort.

Stems prostrate, copiously rooting below, mostly simple with ascending apices; leaves deeply 2-lobed, spreading or somewhat involute when dry, pale or becoming whitish, the lower lobe obliquely ovate or ovate-lanceolate, somewhat falcate, the upper lobe a half smaller, lanceolate, acute; leaf cells rather large, nearly uniform; perianth ovate, with a plicate-laciniate mouth. On rocks. Northern end of Beech Mt. (Rand).

GEOCALYX, Nees.

G. graveolens (Schrad.), Nees.

On rotten stumps. Beech Mt. (Rand).

LOPHOCOLEA, Dumort.

L. bidentata (L.), Dumort.

In rill by roadside, near head of Great Pond (Rand).

CHILOSCYPHUS, Corda.

C. polyanthos (L.), Corda.

On ground. Southern end of Great Pond; Cold Brook (Rand).

Var. **rivularis,** Nees.

On dripping rocks, northern end of Beech Mt. (Rand).

PLAGIOCHILA, Dumort.

P. asplenoides (L.), Dumort.

On wet rocks. Cold Brook; Intervale Brook; northern end of Beech Mt. (Rand).

MYLIA, S. F. Gray.

M. Taylori (Hook.), S. F. Gray.

Among sphagnum, shore of Aunt Bettys Pond (Rand).

M. anomala (Hook.), S. F. Gray.

Differs from *M. Taylori,* of which it may be only a variety, in its rather distant leaves, which are obtuse, acute, or ovate-acuminate on the same stem, and thinner in texture, in its longer perianth, and in its ovate involucral leaves. Among sphagnum, Freeman Heath (E. Faxon).

JUNGERMANNIA, L.

J. Schraderi, Mart.

On ground and old logs; common.

J. pumila, With.

On ground. Woods near Somes Pond (Rand).

J. barbata, Schreb.

On rocks. Western Mt. (Rand). A variety on wet rocks, northern end of Beech Mt. (Rand).

J. attenuata, Lindenb. *J. barbata,* var. *attenuata,* Mart. Gray, Man., 6th ed., 719.

On rocks, northern end of Beech Mt.; on old tree, The Heath, Great Cranberry Isle (Rand).

J. ventricosa, Dicks.

On old log, northern end of Beech Mt. (Rand).

J. excisa, Dicks.

On decaying logs. Jordan Mt.; The Heath, Great Cranberry Isle (Rand).

J. incisa, Schrad.

On the ground and on decaying logs. Southern end of Great Pond (Rand).

J. inflata, Huds.

On ground and rocks ; frequent. Variable. Browns Mt.; Beech Mt.; Sargent Mt. (Rand). A variety on wet rocks, Sargent Mt. (Rand). A form with very small compressed leaves, Robinson Mt. (E. Faxon). An aquatic form, pools on summit of Beech Mt.; on logs in water, Sunken Heath Brook (Rand).

* **J. Marchica,** Nees.

Under leaves (amphigastria) none; stem creeping, radiculose, flexuous, subsimple or with offshoots at the apex, rather thick, soft; leaves semi-vertical, spreading, very lax, subquadrate, repand, entire, pale, bifid with an angular sinus, retrorsely gibbous and with divergent obtuse laciniæ (or more rarely trifid) ; fruit unknown. Nees, Europ. Lebermoose, ii. 77. On *Sphagnum Russowii,* Beech Mt. (Rand). Spec. in herb. C. Warnstorf.

Nees did not include this species in the later work, Synopsis Hepaticorum, nor is it mentioned in other works of importance.

Dr. Warnstorf, however, writes that the discovery is a very important one, since it is the second or third time that this species has been collected. He evidently regards it as a good species.

MARSUPELLA, Dumort.

M. sphacelata (Giesecke), Dumort.
On rocks. Summit of Green Mt. (D. C. Eaton); — Beech Cliff (Rand).

M. emarginata (Ehrh.), Dumort.
Wet rocks; frequent.

M. adusta (Nees), Spruce.
Wet ledges, Green Mt. (E. Faxon).

NARDIA, S. F. Gray.

N. crenulata (Sm.), Lindb.
On decaying wood. The Heath, Great Cranberry Isle (Rand).

FOSSOMBRONIA, Raddi.

F. Dumortieri, Lindb.
Pond shores, on ground. Jordan Pond; Ripples Pond; Great Pond (Rand).

PALLAVICINIA, S. F. Gray.

P. Lyellii (Hook.), S. F. Gray. *Steetzia Lyellii*, Lehm.
Forming mats under water, Sunken Heath Brook (Rand).

PELLIA, Raddi.

P. epiphylla (L.), Nees.
On damp ground, pond shores, brooksides, etc.; common.

ANEURA, Dumort.

A. palmata (Hedw.), Dumort.
Diœcious, generally proliferous, small, opaque; thallus short and narrow; branches linear, palmately divided, usually narrow-

15

ing gradually towards the apex, subacute and scarcely emargi-
nate, biconvex; cells small, rounded, thickened; bracts numerous;
calyptra small, densely verrucose ; antheridia linear.   Lindb.,
Not. pro Fauna et Fl. Fen., xiii. 375. On old tree, West Branch
of Stanley Brook (Rand).

MARCHANTIACEÆ. LIVERWORTS.

MARCHANTIA, L.

M. polymorpha, L.

On the ground, especially after fires; frequent.   Beech Hill;
High Head; Somesville, and elsewhere (Rand).

CONOCEPHALUS, Neck.

C. conicus (L.), Dumort.

On damp ground.   Aunt Mollys Beach (Rand) ; — Bar
Harbor (Kate Furbish).

Class III. THALLOPHYTA.

Division I. CHARACEÆ.

NITELLA, Ag.

N. opaca, Ag.

Ponds and streams; common. Smith Brook, High Head ; Deer Brook; Canada Brook; Jordan Stream (Rand); — Long Pond, Eden (E. Faxon); — at outlet Hadlock Lower Pond (Isaac Holden).

N. flexilis, Ag.

Specimens of a Nitella, probably this species, have been found in several localities. Mouth of Hunters Brook; Somes Stream; Deer Brook (Rand); — Great Pond (Isaac Holden).

Division II. ALGÆ.

List prepared by Frank S. Collins. Plants collected by Frank S. Collins and Isaac Holden.

So far as fresh water algæ are concerned, the following list contains only a few species which have come under the notice of the collectors ; to make even an approximately representative list would take careful collecting and study for years; this must be left for future students. In the representation of marine algæ, the list is more satisfactory, but here also it is undoubtedly far from perfect. For instance, there will be noticed three species which find a place here only on authority of minute fronds observed on other specimens; it is more than likely that there are other species equally deserving of a place, but not equally fortunate in securing it.

The fact that thus far collecting has been done almost entirely in the summer months accounts for the absence of some species, which there is every reason to expect in this locality. Still, allowing for these deficiencies, the list gives a fairly good idea of the marine flora of the Island, — that sub-arctic flora characteristic of the northern New England coast.

The arrangement of species, genera, etc., is based chiefly on Engler and Prantl's "Natürlichen Pflanzenfamilien," which differs somewhat from the standard work of reference for this region, Farlow's "Manual of the Marine Algæ of New England." Where the name here used differs from that employed by Farlow, the latter is given as a synonym; in the case of species and genera not to be found in Farlow, short descriptions have been given, which it is hoped will enable the collector to recognize the plants. Descriptions of the few fresh water algæ will be found in Wolle's "Fresh Water Algæ of the United States."

SUBDIVISION I. RHODOPHYCEÆ.

CORALLINACEÆ.

CORALLINA, L.

C. officinalis, L.

Common in tide pools and below on the shore (Collins); — Sea Wall (Holden).

LITHOTHAMNION, Phil.

L. polymorphum (L.), Aresch.

Common in tide pools (Collins).

L. fasciculatum (Lam.), Aresch.

Occasional on shells, etc. Little Cranberry Isle (Collins). This is the *L. fasciculatum* of Farlow's Manual, p. 182, but it is doubtful if it is the European species of that name. The Lithothamnia of northern Europe have been much studied during the past few years, and it is quite likely that our plant belongs to one of the new species formerly included under *L. fasciculatum ;* but it is impossible to decide without comparison of authentic specimens.

LITHOPHYLLUM, Phil.

L. Lenormandi, Rosanoff. *Melobesia Lenormandi,* Farlow's Manual, 181.

Common in tide pools (Collins).

MELOBESIA, Lamour.

M. pustulata, Lamour.

Occasional on Chondrus, etc. (Collins).

M. Lejolisii, Rosanoff.

A few fronds on Zostera, Little Cranberry Isle (Collins).

SQUAMARIACEÆ.

PEYSSONNELIA, Decne.

P. Dubyi, Crouan.

On shells and stones, near Seal Harbor; not common (Collins).

PETROCELIS, J. Ag.

P. cruenta, J. Ag.

Common in tide pools (Collins); — Greening Island (Holden).

RHIZOPHYLLIDACEÆ.

POLYIDES, Ag.

P. rotundus (Gmelin), Grev.

Occasional in tide pools, near Seal Harbor (Collins) ; — Sea Wall (Holden).

GLŒOSIPHONIACEÆ.

GLŒOSIPHONIA, Carm.

G. capillaris (Huds.), Carm.

Not uncommon. Little Cranberry Isle; near Seal Harbor (Collins).

230

230 FLORA OF MOUNT DESERT.

CERAMIACEÆ.

CERAMIUM, Lyngb.

C. Hooperi, Harv.

Common on overhanging rocks near low-water mark. Near Seal Harbor (Collins); — Sea Wall (Holden).

C. rubrum (Huds.), Ag.

Common everywhere (Collins); — Sea Wall (Holden).

ANTITHAMNION, Naeg.

A. Pylaisæi (Mont.), Farlow. *Callithamnion Pylaisæi,* Farlow's Manual, 123.

On *Ptilota pectinata* (L. R. Boggs).

PLUMARIA, Stack.

P. elegans, Bonnem. *Ptilota elegans,* Farlow's Manual, 133.

On overhanging rocks near low-water mark. Near Seal Harbor (Collins).

PTILOTA, Ag.

P. pectinata (Gunner), Kjellm. *P. serrata,* Farlow's Manual, 133.

Common; cast up from deep water. Seal Harbor, and elsewhere (Collins); — Sea Wall (Holden).

RHODOCHORTON, Naeg.

R. Rothii (Engl. Bot.), Naeg. *Callithamnion Rothii,* Farlow's Manual, 121.

Common on rocks near low-water mark. Seal Harbor, and elsewhere (Collins); — Greening Island (Holden).

R. membranaceum, Magnus.

A minute species, growing in Polyzoa, sponges, etc.; the filaments forming a more or less dense network, sometimes entirely filling the interior of the host, which is then quite noticeable, being of a bright red, instead of the usual yellowish or whitish color. The tetraspores are usually formed outside the host, at the tips of short branches which come out through the host walls. In tubes of Sertularia, etc. Near Seal Harbor (Collins).

RHODOMELACEÆ.

POLYSIPHONIA, Grev.

P. urceolata (Lightf.), Grev.

Not uncommon in tide pools. Little Cranberry Isle; Seal Harbor (Collins).

P. nigrescens (Dillw.), Grev.

One large plant floating at Bracy Cove (Collins).

P. fastigiata (Roth), Grev.

Common on Ascophyllum, all along the shore (Collins); — Somes Sound (Holden).

* P. Olneyi, Harv.

A single specimen on Zostera floating in Somes Sound, in poor condition, but in fruit and unmistakable. Specimen not preserved (Holden).

P. violacea (Roth), Grev.

Somes Sound (Holden).

RHODOMELA, Ag.

R. subfusca (Woodw.), Ag.

Sea Wall (Holden).

DELESSERIACEÆ.

DELESSERIA, Lamour.

D. sinuosa (Good. & Woodw.), Lamour.

Occasional in tide pools, and from deep water (Collins); — Sea Wall (Holden).

D. alata (Huds.), Lamour.

A minute frond of this species on *Ptilota pectinata* (L. R. Boggs).

RHODYMENIACEÆ.

RHODYMENIA, J. Ag.

R. palmata (L.), Grev. DULSE.

Common everywhere on the shore (Collins); — Sea Wall (Holden).

RHODOPHYLLIDACEÆ.

EUTHORA, J. Ag.

E. cristata (L.), J. Ag.

A minute frond of this species on *Ptilota pectinata* (L. R. Boggs).

CYSTOCLONIUM, Kuetz.

C. purpurascens (Huds.), Kuetz.

Not uncommon. Little Cranberry Isle, and elsewhere (Collins).

GIGARTINACEÆ.

AHNFELDTIA, Fries.

A. plicata (Huds.), Fries.

Rather common in tide pools (Collins); — Sea Wall (Holden).

GIGARTINA, Lamour.

G. mamillosa (Good. & Woodw.), J. Ag.

Common near low-water mark (Collins); — Sea Wall (Holden).

CHONDRUS, Stack.

C. crispus (L.), Stack. IRISH MOSS.

Common nearly everywhere on the coast of the Island (Collins).

GELIDIACEÆ.

CHOREOCOLAX, Reinsch.

C. Polysiphoniæ, Reinsch.

This species forms whitish spherical lumps not larger than a pin's head, on the fronds of Polysiphonia, especially at the forkings of the branches. The fronds consist of closely packed, radiating filaments, near the ends of which are formed, in separate individuals, the autheridia, the cystocarps, and the tetraspores. The last, which are cruciately divided, are the commonest form of fruit. Occasional. Seal Harbor, and elsewhere (Collins).

HELMINTHOCLADIACEÆ.

NEMALION, Duby.

N. multifidum, Ag.
Sea Wall (Holden).

CHANTRANSIA, Fries. (*Trentepohlia*, Farlow's Manual, 108.)

C. virgatula (Harv.), Thuret. *T. virgatula*, Farlow's Manual, 109.

On Alaria, etc. Near Seal Harbor (Collins); — Sea Wall (Holden).

C. Daviesii (Engl. Bot.), Thuret. *T. Daviesii*, Farlow's Manual, 109.

On Rhodymenia, Bracy Cove (Collins).

* C. Hermanni (Roth), Kuetz.

On Tuomeya (Holden).

BATRACHOSPERMUM, Roth.

B. vagum, Ag.

Denning Brook (Holden) ; — Deer Brook, Jordan Pond ; Sunken Heath Brook (Rand).

B. pyramidale, Sirdt.

Streamlet, Norwood Cove (Holden); — brook, High Head meadow (Faxon & Rand).

LEMANEACEÆ.

TUOMEYA, Harv.

T. fluviatilis, Harv.

Outlet of Hadlock Lower Pond; Denning Brook (Holden).

LEMANEA, Bory.

L. fucina, Bory, var. rigida, Atkinson.

Hadlock Lower Pond, at outlet (Holden).

PORPHYRACEÆ.

BANGIA, Lyngb.

B. fusco-purpurea (Dillw.), Lyngb.

On rocks. Near Seal Harbor (Collins); — east of Seal Harbor (Holden).

PORPHYRA, Ag.

P. laciniata (Lightf.), Ag.

Sea Wall (Holden). A coarse and dull-colored form on rocks between tide marks ; a smaller and brighter colored form on Fucaceæ, etc. (Collins).

P. miniata, Ag.

A handsome species, differing from *P. laciniata* in having two layers of cells instead of one in the greater part of the frond, even in the vegetative condition. The fronds are more gelatinous and somewhat thicker, and adhere firmly to paper when dried. It is a species of deeper water than *P. laciniata*, and grows on other algæ, rather than on rocks or woodwork. Floating near Seal Harbor (Collins).

Genera of Doubtful Affinity.

HILDENBRANDTIA, Nardo.

H. prototypus, Nardo. *H. rosea,* Farlow's Manual, 116.
Very common on rocks in tide pools, etc. (Collins).

HALOSACCION, Kuetz.

H. ramentaceum (L.), Ag.
Common in lower tide pools (Collins); — Sea Wall (Holden).

SUBDIVISION II. PHÆOPHYCEÆ.

FUCACEÆ.

ASCOPHYLLUM, Stack. ROCKWEED.

A. nodosum (L.), Le Jolis.
Very common everywhere on the coast (Collins).

FUCUS, L. ROCKWEED.

F. filiformis, Gmelin.
In upper tide pools. Near Seal Harbor (Collins).

F. edentatus, De la Pyl. *F. furcatus,* Farlow's Manual, 102.
Near Life Saving Station, Little Cranberry Isle (Collins).

F. evanescens, Ag.
Common all along the shore (Collins).

F. vesiculosus, L.
Very common everywhere on the coast (Collins); — Somes
Sound (Holden). Occurring in numerous forms, among them:

Var. **laterifructus,** Grev.
With the type (Collins).

F. platycarpus, Thuret.
Resembling *F. edentatus,* but with the branching somewhat
lateral, rather than regularly forked, the conceptacles shorter

and rounded, and margined with the unchanged membrane of the frond. It grows nearer high-water mark than either *F. evanescens* or *F. edentatus.* Somes Sound, on rocks near high-water mark (Holden).

LAMINARIACEÆ.

ALARIA, Grev.

A. esculenta (L.), Grev.

Rather common in tide pools and below (Collins); — Sea Wall (Holden).

A. Pylaii (Bory), J. Ag. *A. esculenta,* var. *latifolia,* Farlow's Manual, 97.

Rather common (Collins). Regarded as a distinct species by most authors, and in its extreme forms quite different in appearance from the preceding species.

AGARUM, Bory.

A. Turneri, Post. & Rupr.

Common near and below low-water mark (Collins); — Sea Wall (Holden).

SACCORHIZA, De la Pyl.

S. dermatodea, De la Pyl.

Rather common in lower tide pools (Collins); — Sea Wall (Holden).

LAMINARIA, Lamour. DEVIL'S APRON.

L. saccharina (L.), Lamour.

Common in lower tide pools and below (Collins).

L. longicruris, De la Pyl.

Occasional. From deep water (Collins); — floating, Somes Sound (Holden).

L. platymeris, De la Pyl.

Not uncommon. From deep water (Collins).

L. digitata, Lamour.

Sea Wall (Holden).

Var. **ensifolia,** Le Jolis.

A rather small form, with numerous narrow linear segments, and rather slender, rounded stipe; growing mostly in shallow water. In tide pools, Little Cranberry Isle, and elsewhere (Collins).

CHORDA, Stack.

C. filum (L.), Stack.

Rather common at Little Cranberry Isle (Collins).

RALFSIACEÆ.

RALFSIA, Berk.

R. clavata (Carm.), Crouan.

On woodwork and shells near Seal Harbor (Collins).

R. pusilla (Stroemf.), Holmes & Batters.

A minute plant, forming a dark brown or black coating on the filaments of Chætomorpha; somewhat like *R. clavata,* but a much smaller plant in all its parts ; the habitat quite distinct, — *R. pusilla* growing only on algæ, *R. clavata* on any hard lifeless substance indiscriminately. On *Chætomorpha Melagonium,* Sea Wall (Holden).

R. verrucosa (Aresch.), J. Ag.

Common all along the shore (Collins).

R. deusta, J. Ag.

At and below low-water mark, near Seal Harbor (Collins) ; — tide pool, Sea Wall (Holden).

CHORDARIACEÆ.

CHORDARIA, Ag.

C. flagelliformis (Fl. Dan.), Ag.

Common all along the shore (Collins) ; — Somes Sound (Holden).

238

LEATHESIA, S. F. Gray.

L. difformis (L.), Aresch.

Common in lower tide pools, all along the shore (Collins); — Sea Wall (Holden).

CASTAGNEA, Derb. & Sol.

C. virescens (Carm.), Thuret.

In tide pools. Near Life Saving Station, Little Cranberry Isle; near Seal Harbor (Collins).

MYRIONEMA, Grev.

M. strangulans, Grev. *M. vulgare,* Farlow's Manual, 79.

On various algæ. Little Cranberry Isle; near Seal Harbor (Collins)

ELACHISTEACEÆ.

ELACHISTEA, Duby.

E. fucicola (Velley), Fries.

On Fucus. Very common all along the shore (Collins); — Southwest Harbor (Holden).

E. lubrica, Rupr.

On Halosaccion and Ascophyllum. Near Seal Harbor (Collins).

DESMARESTIACEÆ.

DESMARESTIA, Lamour.

D. aculeata (L.), Lamour.

From deep water. Little Cranberry Isle (Collins); — Southwest Harbor (Holden).

D. viridis (Fl. Dan.), Lamour.

From deep water. Little Cranberry Isle; Seal Harbor (Collins).

DICTYOSIPHONACEÆ.

DICTYOSIPHON, Grev.

D. hippuroides (Lyngb.), Aresch.

Common. Little Cranberry Isle; near Seal Harbor, and elsewhere (Collins); — " Mt. Desert " (Holden).

D. fœniculaceus (Huds.), Grev.

Common with the last (Collins);—Southwest Harbor (Holden).

ENCŒLIACEÆ.

ASPEROCOCCUS, Lamour.

A. echinatus (Mert.), Grev.

On Fucaceæ. Near Seal Harbor (Collins).

PHYLLITIS, Kuetz.

P. fascia (Fl. Dan.), Kuetz.

Common in tide pools along the shore (Collins); — Sea Wall (Holden).

Var. **cæspitosa** (J. Ag.), Farlow.

With the type, but not so common (Collins).

SCYTOSIPHON, Ag.

S. lomentarius (Lyngb.), J. Ag.

Very common everywhere on the coast (Collins).

PUNCTARIA, Grev.

P. latifolia, Grev.

Occasional at Seal Harbor (Collins).

DESMOTRICHUM, Kuetz.

Similar to Punctaria, but the frond consists of one row or a few rows of cells; the plurilocular sporangia either immersed, sessile, or on short stalks.

D. undulatum (J. Ag.), Reinke. *Punctaria latifolia,* var. *Zosteræ,* Farlow's Manual, 64, at least in part.

On Zostera. Little Cranberry Isle (Collins); — Somes Sound (Holden).

SPHACELARIACEÆ.

SPHACELARIA, Lyngb.

S. radicans (Dillw.), Ag.

Common on rocks near Seal Harbor (Collins).

ECTOCARPACEÆ.

ECTOCARPUS, Lyngb.

E. confervoides (Roth), Le Jolis.

Common on Chorda, Zostera, etc. Seal Harbor, and elsewhere (Collins); — Southwest Harbor (Holden).

Var. **siliculosus,** Kjellm.

On Zostera. Little Cranberry Isle (Collins); — Southwest Harbor (Holden).

E. fasciculatus, Harv.

Common on Rhodymenia, Laminaria, etc. Seal Harbor, and elsewhere (Collins); — Sea Wall (Holden).

ASCOCYCLUS, Magnus.

Resembles Myrionema, but the sporangia terminate upright filaments or their branches, instead of rising directly from the basal layer.

A. orbicularis (J. Ag.), Magnus.

Basal stratum of one layer of cells, from which arise colorless hairs, unicellular saccate paraphyses, and shortly stipitate plurilocular sporangia, usually of a single series of cells. On Zostera, Little Cranberry Isle (Collins).

PYLAIELLA, Bory.

P. littoralis (L.), Kjellm. *Ectocarpus littoralis,* Farlow's Manual, 73.

Common on Fucaceæ, etc. (Collins).

Var. robustus, Farlow.

Floating, near Seal Harbor (Collins).

SUBDIVISION III. CHLOROPHYCEÆ.

CONJUGATÆ.

SPIROGYRA, Link.

* S. majuscula, Kuetz.

In fresh water (Holden).

PROTOCOCCACEÆ.

PROTOCOCCUS, Ag.

* P. viridis, Ag.

Shaded places; rocks, trees, fences, etc.; common. (Holden.)

VAUCHERIACEÆ.

VAUCHERIA, DC.

V. Thuretii, Woronin.

Very common in lagoon, Little Cranberry Isle (Collins); — shore west of Bracy Cove, with Microcoleus, etc. (Holden).

V. litorea, Hoffm. Bang.

Rather common at Little Cranberry Isle (Collins).

16

VALONIACEÆ.

CODIOLUM, A. Br.

C. longipes, Foslie.

Fronds proportionally longer and slenderer than in typical *C. gregarium.* Common on rocks near Seal Harbor (Collins); — on rocks between tide marks, Sea Wall (Holden).

C. gregarium, A. Br.

A few plants of the typical form of this species have been found mixed with *C. longipes* and with transitional forms. It is doubtful whether the two species are distinct. If they are not, *C. gregarium* as a specific name has the priority. Seal Harbor (Collins).

GOMONTIACEÆ.

GOMONTIA, Born. & Fl.

Fronds branching; individual cells transformed into large round, oval, or clavate sporangia, which at length separate from the frond, and develop zoospores and resting spores.

G. polyrhiza (Lagerh.), Born. & Fl.

Filaments .004–.008 mm. diam.; sporangia .03–.04 mm. diam. Appears as a grass-green stain on dead shells. Seal Harbor (Collins).

CONFERVACEÆ.

CLADOPHORA, Kuetz.

C. arcta (Dillw.), Kuetz.

Very common in tide pools (Collins); — Somes Sound (Holden).

Forma **centralis.**

Probably merely an older stage. Sea Wall (Holden).

C. lanosa (Roth), Kuetz.

In tide pools near Seal Harbor (Collins). Apparently not very common.

C. glaucescens (Griff.), Harv.

In tide pools, Little Cranberry Isle (Collins).

C. lætevirens (Dillw.), Harv.

Common in tide pools. Little Cranberry Isle; near Seal Harbor (Collins).

C. gracilis (Griff.), Kuetz.

In lower tide pools. Near Seal Harbor (Collins); — Somes Sound (Holden).

C. expansa, Kuetz.

In upper tide pools, Seal Harbor; very common in lagoon, Little Cranberry Isle (Collins); — Long Pond (Holden).

C. rupestris (L.), Kuetz.

On rocks in tide pools, Sea Wall (Holden).

C. flexuosa (Griff.), Harv.

Sea Wall (Holden).

RHIZOCLONIUM, Kuetz.

R. riparium (Roth), Harv.

Very common all along the shore (Collins); — Norwood Cove (Holden).

R. tortuosum, Kuetz.

Common in tide pools (Collins); — Sea Wall (Holden).

CHÆTOMORPHA, Kuetz.

C. Picquotiana (Mont.), Kuetz.

From deep water, Seal Harbor (Collins).

C. Melagonium (Web. & Mohr), Kuetz.

Sea Wall (Holden). A form found on pebble in lagoon, Little Cranberry Isle (Collins), probably belongs to this species, though only about half the usual diameter.

BULBOCOLEON, Prings.

B. piliferum, Prings.

In fronds of *Castagnea virescens* from tide pools, Seal Harbor (Collins).

ULOTHRIX, Kuetz.

U. flacca (Dillw.), Thuret.

On piles of wharves and on rocks. Seal Harbor; Southwest Harbor (Collins).

U. isogona (Sm.), Thuret.

On stones near Life Saving Station, Little Cranberry Isle (Collins).

* U. zonata, Kuetz.

In fresh water (Holden).

CONFERVA, Link.

* C. affinis, Kuetz.

In fresh water (Holden).

ULVACEÆ.

TETRANEMA, Aresch.

Frond consisting at first of a single series of cells, subsequently of two (or four?) series arranged symmetrically.

T. percursum (Ag.), Aresch.

Forming light green or yellowish masses in warm upper pools, usually mixed with various species of Enteromorpha, Cladophora, etc. The frond consists at first of a single conferva-like filament, which soon divides into two series of cells, set side by side, the cells, usually a little longer than broad, being set symmetrically, the cross walls exactly opposite. The filaments are unbranched, usually .01 to .016 mm. diam. In upper tide pools. Near Seal Harbor (Collins).

CAPSOSIPHON, Gobi.

Frond tubular, formed of longitudinally arranged gelatinous cells, which divide in two directions, the walls of the mother cells persistent for a time, as in Glœocapsa.

C. aureolus (Ag.), Gobi.

Resembles a slender unbranched Enteromorpha, but the cells look like Glœocapsa, being rounded, dividing by twos or fours, the mother cell-wall showing somewhat after the division. The longitudinal arrangement of cells is very distinct in the filament, a little pressure on the cover glass under the microscope often dividing the frond for quite a distance up and down. This species seems to prefer localities where it is exposed alternately to fresh and to salt water. Common on stones in brook flowing from Long Pond through the beach (Collins).

ENTEROMORPHA, Link.

E. Linza (L.), J. Ag. *Ulva enteromorpha*, var. *lanceolata,* Farlow's Manual, 43.

Very common all along the shore (Collins).

E. intestinalis (L.), Link. *Ulva enteromorpha*, var. *intestinalis*, Farlow's Manual, 43.

Very common along the shore (Collins); — Norwood Cove (Holden).

E. micrococca, Kuetz.

Resembles *E. intestinalis*, but is a smaller plant every way, rarely if ever an inch in length. The cells are very small, .004–.005 mm. diameter. It usually grows in dense masses on cliffs between tidemarks, in places always wet by streams from above or by dripping water. Cliff near Seal Harbor (Collins).

E. compressa (L.), Grev. *Ulva enteromorpha*, var. *compressa,* Farlow's Manual, 43.

Common along the shore (Collins).

E. clathrata (Roth), J. Ag.　*Ulva clathrata,* Farlow's Manual, 44.

In tide pools (Collins); — Somes Sound (Holden).

E erecta (Lyngb.), J. Ag.

Resembles *E. clathrata,* but the cells have more compact and opaque contents, and are less regularly arranged in series. The smaller branches seem articulate, not conferva-like, as in *E. Hopkirkii,* but after the manner of a Polysiphonia, several cells side by side of the same height.　Shore, Little Cranberry Isle (Collins).

E. Hopkirkii, McCalla.　*Ulva Hopkirkii,* Farlow's Manual, 44.

In tide pools, on *Cladophora glaucescens,* Little Cranberry Isle (Collins).

MONOSTROMA, Thuret.

* M. Blyttii (Aresch.), Wittr.

Seal Harbor (Elisa W. Redfield).

ULVA, L.　SEA LETTUCE.

U. lactuca (L.), Le Jolis.

Occurring in two forms.

Var. rigida (Ag.), Le Jolis.

Common in tide pools and below (Collins).

Var. lactuca, Le Jolis.

Common with the last (Collins).

SUBDIVISION IV.　SCHIZOPHYCEÆ.

HORMOGONEÆ.

CALOTHRIX, Ag.

C. scopulorum (Web. & Mohr), Ag.

Very common on rocks.　Seal Harbor; Little Cranberry Isle (Collins); — Sea Wall (Holden).

C. pulvinata (Mert.), Ag.

On piles of bridge, outlet of Long Pond (Collins). Rare; the most northern station for this species yet reported.

RIVULARIA, Roth.

R. atra, Roth.

Common in upper tide pools (Collins); — Sea Wall (Holden).

R. nitida, Ag. *R. plicata,* Farlow's Manual, 38.

On woodwork; rare (Collins).

MASTIGOCOLEUS, Lagerh.

Trichome of a single series of cells, blue-green or yellowish, .006–.01 mm. diam., branching, the branches sometimes of uniform diameter, sometimes tapering to a fine hair. Heterocysts terminal or lateral; spores unknown. Appears as a blue-green stain on the surface of the shell.

M. testarum, Lagerh.

Growing in the substance of dead shells, in company with *Hyella cæspitosa,* etc. Seal Harbor (Collins). The only species of the genus.

STIGONEMA, Ag.

S. mamillosum, Ag.

On rocks in outlet of Hadlock Lower Pond (Holden).

MICROCOLEUS, Desmaz.

M. chthonoplastes (Fl. Dan.), Thuret.

Very common in lagoon, Little Cranberry Isle (Collins); — shore west of Bracy Cove (Holden).

LYNGBYA, Ag.

L. æstuarii (Juerg.), Liebm.

Very common in lagoon, Little Cranberry Isle ; occasional near Seal Harbor (Collins).

L. semi-plena (Ag.), J. Ag.

Filaments .006–.012 mm. diam., sheath usually thin and delicate; articulations ¼–½ as long as broad; color of stratum dark green to olive-yellow. Not common. Shore near Seal Harbor (Collins).

L. lutea, Gomont. *L. tenerrima*, Farlow's Manual, 35.

Near outlet of Long Pond, among Calothrix, etc. (Collins).

PHORMIDIUM, Kuetz.

Similar to Oscillatoria, but the filaments are included in a general mucilaginous layer.

P. fragile, Gomont.

Trichomes bright green, somewhat flexuous, moniliform, attenuate at the apex, cells subquadrate, .0012–.0023 mm. diam. Forming a dull green gelatinous stratum on woodwork. Near Seal Harbor (Collins).

OSCILLATORIA, Vauch.

O. subuliformis, Kuetz.?

On rocks near Seal Harbor (Collins). A plant agreeing with the description of *Oscillaria subuliformis*, Harv., in Farlow's Manual, 33. According to Gomont, it is doubtful if Harvey's plant is the same as Kuetzing's, so that all that can be said in this case is that the plant agrees with Farlow's description.

* **O. limosa,** Ag.

In fresh water (Holden).

* **O. amphibia,** Ag.

In fresh water (Holden).

SPIRULINA, Turpin.

S. subsalsa, Oersted. *S. tenuissima*, Farlow's Manual, 31.

On rocks on shore near Seal Harbor (Collins).

CHROOCOCCACEÆ.

CHROOCOCCUS, Naeg.

C. turgidus, Naeg.
Common among various algæ in lagoon, Little Cranberry Isle (Collins).

GLŒOCAPSA, Kuetz.

G. crepidinum, Thuret.
Common on rocks, etc., near high-water mark (Collins).

POLYCYSTIS, Kuetz.

P. elabens, Kuetz.
Among small algæ, Seal Harbor (Collins).

P. pallida (Kuetz.), Farlow.
Among small algæ, Seal Harbor (Collins).

CHAMÆSIPHONACEÆ.

DERMOCARPA, Crouan.

D. prasina (Reinsch), Born. *Sphœnosiphon smaragdinus*, Farlow's Manual, 61.
On *Polysiphonia fastigiata.* Near Seal Harbor (Collins).

HYELLA, Born. & Fl.

Filaments .004–.012 mm. diam., forming a horizontal net-like layer, from which arise vertical filaments. Cells not close together, as in Lyngbya, often Chroococcus-like. Forms grayish stains on shells.

H. cæspitosa, Born. & Fl.
The only species of the genus. Growing in the substance of dead shells, in company with *Mastigocoleus testarum*, etc. Seal Harbor (Collins).

DIVISION III. LICHENES.

List contributed by Dr. John W. Eckfeldt, and based on collections made by him, by Miss Mary L. Wilson, and by others.
Plants determined by Dr. Eckfeldt, Miss Wilson, and Miss Clara
E. Cummings. As all collections have hitherto been made either
within a short period of time, or under adverse circumstances,
doubtless a more thorough examination of the Island would
extend the list very greatly. In the case of Dr. Eckfeldt's
collections the names of the special stations have in most cases
unfortunately been lost.

TRIBE I. **PARMELIACEI.**

USNEEI.

RAMALINA, Ach.

R. calicaris (L.), Fries.

On trees. (Eckfeldt); — Seal Harbor (Wilson).

* Var. **fraxinea,** Fries.

(Eckfeldt.)

Var. **fastigiata,** Fries.

Seal Harbor (Wilson).

Var. **canaliculata,** Fries.

(Eckfeldt); — on fir trees, Seal Harbor (Wilson).

Var. **farinacea,** Schaer.

(Eckfeldt); — Seal Harbor (Wilson).

* **R. intermedia,** Delis.

Thallus flat, attenuated and mostly divided, the margins frequently sorediate. Apothecia pale yellow, terminal, subtended
by the elongated forked extremity of the lacinia. On shrubs and
branches (Eckfeldt). Frequently mistaken for the preceding
variety.

R. pusilla, Prev.

On apple trees, Seal Harbor (Wilson).

* **R. pollinaria,** Ach.

Small and imperfect specimen, but distinct (Eckfeldt).

* **R. polymorpha,** Ach.

On rocks in small patches (Eckfeldt).

CETRARIA, Ach.

C. aculeata (Schreb.), Fries.

Green Mt. (Wilson, T. G. White); — Browns Mt. (Rand).

* **C. Islandica** (L.), Ach.

On the ground. (Eckfeldt.)

* Var. **Delisæi** (Bor.), Nyl.

(Eckfeldt.)

C. cucullata (Bell), Ach.

On the ground. (Eckfeldt.)

C. nivalis (L.), Ach.

On the ground. (Eckfeldt.)

C. aleurites (Ach.), Th. Fries.

Generally sterile. On trees and dead wood (Eckfeldt); — on fence rails, Seal Harbor (Wilson).

C. Fahlunensis (L.), Schaer.

On rocks. (Eckfeldt); — Seal Harbor (Wilson).

C. ciliaris, Ach.

On trees and fence rails. (Eckfeldt); — Seal Harbor (Wilson); — Somesville (E. Faxon).

C. lacunosa, Ach.

On trees. (Eckfeldt); — Seal Harbor (Wilson).

* Var. **stenophylla,** Tuck.

(Eckfeldt.)

C. glauca (L.), Ach.

On trees and rocks. (Eckfeldt); — Seal Harbor (Wilson, Redfield).

C. Oakesiana, Tuck.

On trees and rocks. (Eckfeldt); — Seal Harbor (Wilson).

* C. aurescens, Tuck.

On old fences. Not common (Eckfeldt).

* C. juniperina (L.), Ach.

On trees. (Eckfeldt.)

Var. terrestris, Schaer.

On the ground. (Eckfeldt.)

Var. Pinastri, Ach.

On trees. (Eckfeldt); — Seal Harbor (Wilson).

EVERNIA, Ach.

* E. vulpina (L.), Ach.

On trees and fence rails. Sparingly found in a degenerate condition (Eckfeldt).

* E. furfuracea (L.), Mann.

On trees. (Eckfeldt.)

Var. Cladonia, Tuck.

Seal Harbor (Eckfeldt).

E. Prunastri (L.), Ach.

On trees. Seal Harbor (Eckfeldt, Wilson).

USNEA, Ach.

U. barbata (L.), Fries.

On old trees. Common (Eckfeldt, Redfield, Rand).

Var. florida, Fries.

On trees. Seal Harbor (Wilson); — (Eckfeldt).

* Var. hirta, Fries.

On branches of trees. (Eckfeldt.)

* Var. rubiginea, Mx.

On branches, mingled with var. *hirta* (Eckfeldt).

Var. **ceratina**, Schaer.
On branches of trees. (Eckfeldt.)

Var. **dasypoga**, Fries.
Seal Harbor (Wilson).

Var. **plicata**, Fries.
On trees. (Eckfeldt, Wilson, Redfield.)

U. **trichodea**, Ach.
On trees. Southwest Harbor (Eckfeldt).

ALECTORIA, Ach.

A. **jubata**, L.
On old fence rails and trees. (Eckfeldt);—Sunken Heath (Rand).

* Var. **bicolor**, Fries.
On rails and branches. (Eckfeldt.)

Var. **chalybeiformis**, Ach.
(Eckfeldt);—Seal Harbor (Wilson).

Var. **implexa**, Fries.
On fir trees. Northeast Harbor (Wilson);—Great Cranberry Isle (Rand).

A. **ochroleuca** (Ehrh.), Nyl.
On branches of trees. (Eckfeldt.)

* Var. **rigida**, Fries.
On the ground. Seal Harbor (Eckfeldt).

* Var. **osteina**, Nyl.
(Eckfeldt.)

Var. **nigricans**, Ach.
On the ground. (Eckfeldt.)

* Var. **sarmentosa**, Nyl.
More common than the preceding (Eckfeldt).

PARMELIEI.

THELOSCHISTES, Norm.

* **T. chrysophthalmus** (L.), Norm.

On shrubs. (Eckfeldt.)

T. parietinus (L.), Norm.

On rocks. Eastern coast (Eckfeldt, Wilson, Redfield). On dead trees. Somesville (Rand).

T. polycarpus, Ehrh.

On trees and dead wood. (Eckfeldt) ; — Seal Harbor (Wilson).

T. lychneus, Nyl.

On trees and rocks. (Eckfeldt); — Seal Harbor (Wilson).

* **T. concolor**, Dicks.

On rough barks. (Eckfeldt.)

PARMELIA, Ach.

P. perlata (L.), Ach.

(Eckfeldt); — Seal Harbor (Wilson).

* **P. perforata** (Jacq.), Ach.

On tree trunks. (Eckfeldt.)

* **P. cetrata**, Ach.

On rocks. Seal Harbor (Eckfeldt).

P. crinita, Ach.

On rocks. Seal Harbor (Wilson).

P. aurulenta, Tuck.

On tree trunks and rocks. (Eckfeldt.)

P. tiliacea (Hoffm.), Floerke.

On trees and stones. (Eckfeldt); — Seal Harbor (Wilson).

P. Borreri, Turn.

On rocks and trees. Seal Harbor (Wilson).

* Var. **rudecta**, Tuck.

More frequently seen on tree trunks (Eckfeldt).

P. saxatilis (L.), Fries.

On rocks and trees. (Eckfeldt); — Seal Harbor (Wilson, Redfield).

Var. **sulcata**, Nyl.

Seal Harbor (Eckfeldt).

Var. **omphalodes**, Fries.

On rocks. Seal Harbor (Wilson, Eckfeldt).

P. physodes (L.), Ach.

On rocks. (Eckfeldt); — Seal Harbor (Wilson).

* Var. **obscurata**, Ach.

On rocks. Seal Harbor (Eckfeldt).

* Var. **vittata**, Ach.

On trees. Infertile (Eckfeldt).

* **P. encausta** (Sm.), Nyl.

On rocks. (Eckfeldt.)

P. pertusa (Schrk.), Schaer.

On trees and rocks. (Eckfeldt); — Seal Harbor (Wilson).

P. colpodes (Ach.), Nyl.

On tree trunks. (Eckfeldt, Wilson.)

P. olivacea (L.), Ach.

On trees. (Eckfeldt); — Seal Harbor (Wilson); — Southwest Harbor (M. L. Fernald).

* Var. **aspidota**, Ach.

(Eckfeldt.)

Var. **sorediata** (Ach.), Nyl.

On rocks. Seal Harbor (Wilson).

P. stygia (L.), Ach.

On rocks. Seal Harbor (Wilson).

* P. lanata (L.), Wallr.
   On rocks. (Eckfeldt.)

P. caperata (L.), Ach.
   On trees and decorticated wood. (Eckfeldt, Wilson.)

P. conspersa (Ehrh.), Ach.
   On rocks. (Eckfeldt); — Seal Harbor (Wilson).

P. centrifuga (L.), Ach.
   On rocks. Seal Harbor (Wilson);—"Mt. Desert" (Tuckerman).

P. incurva (Pers.), Fries.
   On rocks. " Mt. Desert" (Tuckerman) ; — Seal Harbor (Wilson).

P. ambigua (Wulf.), Ach.
   On dead wood, etc. (Eckfeldt); — Seal Harbor (Wilson).

* Var. albescens, Wahl.
   On rocks. (Eckfeldt.)

PHYSCIA, DC.

P. speciosa (Wulf.), Nyl.
   On shaded rocks. Seal Harbor (Wilson).

* P. hypoleuca (Muhl.), Tuck.
   On trees. (Eckfeldt.)

* P. aquila (Ach.), Nyl., var. detonsa, Tuck.
   On rocks, etc. (Eckfeldt.)

* P. pulverulenta (Schreb.), Nyl., var. leucoleiptes, Tuck.
   (Eckfeldt.)

P. stellaris, L.
   On trees and rocks. (Eckfeldt); — Seal Harbor (Wilson).

* P. astroidea (Fries), Nyl.
   On trees. (Eckfeldt.)

* P. tribacia (Ach.), Tuck.
   On old tree trunks. (Eckfeldt, Wilson.)

P. hispida (Schreb.), Tuck.
On trees.   (Eckfeldt); — Seal Harbor (Wilson).

P. cæsia (Hoffm.), Nyl.
On rocks.   Seal Harbor (Wilson).

P. obscura (Ehrh.), Nyl.
On trees, etc.   Seal Harbor (Wilson).

* Var. **endochrysea**, Nyl.
At the base of trees (Eckfeldt).

P. adglutinata (Floerke), Nyl., var. **pyrithrocardia**, Mueller.
On trees.   Seal Harbor (Wilson).

PYXINE, Fries.

P. sorediata, Fries.
On rocks.   Seal Harbor (Wilson).

UMBILICARIEI.

UMBILICARIA, Hoffm.

* **U. cylindrica** (L.), Delis.
On rocks.   (Eckfeldt.)

U. polyphylla (L.), Hoffm.
On rocks.   "Mt. Desert" (Tuckerman, Eckfeldt) ; — Seal Harbor (Wilson).

U. flocculosa, Hoffm.
On rocks.   "Mt. Desert" (Tuckerman) ; — Seal Harbor (Wilson).

U. Muhlenbergii (Ach.), Tuck.
On rocks.   (Eckfeldt) ; — Seal Harbor (Redfield).

* **U. hirsuta** (Ach.), Stenh.
On rocks.   (Eckfeldt.)

* **U. vellea** (L.), Nyl.
On rocks.   (Eckfeldt.)

17

U. Dillenii, Tuck.

On rocks. Seal Harbor (Redfield).

U. pustulata (L.), Hoffm., var. **papulosa,** Tuck.

On rocks. (Eckfeldt); — Seal Harbor (Wilson); — Browns Mt. (Rand).

PELTIGEREI.

STICTA, Schreb.

S. amplissima (Scop.), Mass.

On trees and rocks. (Eckfeldt, Wilson, Redfield.)

S. pulmonaria (L.), Ach.

On trees and rocks. (Eckfeldt, Wilson); — Jordan Pond trail, Seal Harbor (Redfield).

S. fuliginosa (Dicks.), Ach.

On rocks. Woods, Seal Harbor (T. G. White).

S. crocata (L.), Ach.

On trees and rocks. Seal Harbor (Wilson, Redfield); — Beech Mt. (E. Faxon).

S. scrobiculata (Scop.), Ach.

On trees and rocks. (Eckfeldt); — Seal Harbor (Wilson, Redfield); — Beech Cliff (E. Faxon).

NEPHROMA, Ach.

* N. arcticum (L.), Fries.

On the ground (Eckfeldt).

N. tomentosum (Hoffm.), Koerber.

On rocks. (Eckfeldt); — Seal Harbor (Wilson, Redfield).

N. Helveticum, Ach.

On rocks. Seal Harbor (Wilson).

* N. lævigatum, Ach.

(Eckfeldt.)

Var. parile, Nyl.

On trees and rocks. Seal Harbor (Wilson).

CATALOGUE OF PLANTS. 259

PELTIGERA, Willd.

* P venosa (L.), Hoffm.
On the ground. (Eckfeldt.)

P. aphthosa (L.), Hoffm.
On the ground and on rocks. (Eckfeldt); — Hadlock Valley; Browns Mt. Notch (Redfield); — Cold Brook (Rand); — Southwest Harbor (M. L. Fernald).

P. horizontalis (L.), Hoffm.
On the ground and on trees. (Eckfeldt, Wilson); — Beech Cliff (E. Faxon); — Sutton Island (Redfield); — Southwest Harbor (M. L. Fernald).

P. polydactyla (Neck.), Hoffm.
On the ground. (Eckfeldt); — Sutton Island (Redfield); — Somesville; Northwest Arm woods (Rand).

P. rufescens (Neck.), Hoffm.
On rocks, etc. (Eckfeldt); — Seal Harbor (Wilson).

P. canina (L.), Hoffm.
On the ground, etc. (Eckfeldt); — Seal Harbor (Redfield).

* Var. spongiosa, Tuck.
(Wilson.)

* Var. membranacea (Ach.), Nyl.
On moss (Eckfeldt).

SOLORINA, Ach.

* S. saccata (L.), Ach.
On the ground. (Eckfeldt.)

PANNARIEI.

HEPPIA, Naeg.

* H. Despreauxii (Mont.), Tuck.
On the earth. Southwest Harbor (Eckfeldt).

PANNARIA, Delis.

P. lanuginosa (Ach.), Koerber.

On rocks. (Eckfeldt); — caves, Barr Hill (Redfield).

* **P. hypnorum** (Hoffm.), Koerber.

On tree trunks and on the earth. (Eckfeldt.)

* **P. granatina**, Sommerf.

On rocks. (Eckfeldt.)

P. rubiginosa (Thunb.), Delis.

On rocks. Seal Harbor (Wilson).

P. leucosticta, Tuck.

On rocks among dead moss (Eckfeldt).

P. brunnea (Swz.), Mass.

On the ground. Crystal Cove, High Head (Rand).

P. plumbea (Lightf.), Delis.

On old oak. Newport Mt. (Tuckerman). On rocks. (Wilson, Eckfeldt.)

P. nigra (Huds.), Nyl.

On rocks. (Eckfeldt.)

COLLEMEI.

OMPHALARIA, Dur. & Mont.

O. phyllisca (Wahl.), Tuck.

Frequent upon rocks, Seal Harbor (Wilson).

COLLEMA, Hoffm.

C. leptaleum, Tuck.

On tree trunks. Seal Harbor (Wilson).

C. flaccidum, Ach.

On old trees and stumps. Seal Harbor (Wilson); — Stanley Brook (Redfield).

C. nigrescens (Huds.), Ach.

On tree trunks. (Eckfeldt.)

LEPTOGIUM, Fries.

L. pulchellum (Ach.), Nyl.

On tree trunks, etc. (Eckfeldt.)

L. tremelloides (L. f.), Fries.

On rocks and tree trunks. (Eckfeldt); — Seal Harbor (Wilson); — Beech Cliff (E. Faxon).

* **L. chloromelum** (Swz.), Nyl.

On old tree trunks. (Eckfeldt.)

LECANOREI.

PLACODIUM, DC.

P. elegans (Link), DC.

On rocks. (Eckfeldt); — Seal Harbor (T. G. White).

* **P. murorum** (Hoffm.), DC.

On rocks. (Eckfeldt.)

* **P. cinnabarinum** (Ach.), Anz.

On rocks. (Eckfeldt.)

* **P. citrinum** (Hoffm.), Leight.

On old mortar. (Eckfeldt.)

P. aurantiacum (Lightf.), Naeg. & Hepp.

On trees, dead wood, etc. (Eckfeldt);—Seal Harbor (Redfield).

* **P. cerinum** (Hedw.), Naeg. & Hepp.

On trees, etc. (Eckfeldt, Wilson.)

Var. **pyracea**, Nyl.

(Eckfeldt.)

* **P. ferrugineum** (Huds.), Hepp, var. **discolor**, Willey.

On oak trees. "Mt. Desert" (Tuckerman).

* **P. camptidium**, Tuck.

On smooth tree trunks. Southwest Harbor (Eckfeldt).

P. vitellinum (Ehrh.), Naeg. & Hepp.

On dead wood. Seal Harbor (Wilson).

LECANORA, Ach.

* L. rubina (Vill.), Ach.
On rocks. (Eckfeldt.)

L. muralis (Schreb.), Schaer.
On rocks. (Eckfeldt, Wilson.)

L. pallida (Schreb.), Schaer.
On trees. (Eckfeldt); — Seal Harbor (Wilson).

* Var. cancriformis, Tuck.
On tree trunks. (Eckfeldt.)

Var. angulosa, Hoffm.
On trees. Seal Harbor (Wilson).

* L. sordida (Pers.), Th. Fries.
On rocks. (Eckfeldt.)

L. subfusca (L.), Ach.
On rocks and wood. (Eckfeldt, Wilson, Redfield.)

* Var. allophana, Ach.
(Eckfeldt.)

* Var. distans, Ach.
(Eckfeldt.)

* L. Hageni, Ach.
On old rails and houses. (Eckfeldt.)

L. varia (Ehrh.), Nyl.
On bark, stones, etc. (Eckfeldt); — Seal Harbor (Wilson).

Var. polytropa, Nyl.
On wood. Jordan Pond (Wilson).

* Var. sæpincola, Fries.
On dead wood. (Eckfeldt.)

* L. ventosa (L.), Ach.
On rocks. (Eckfeldt.)

L. elatina, Ach., var. **ochrophæa,** Tuck.

On bark and dead wood. (Eckfeldt); — Seal Harbor (Wilson).

L. pallescens (L.), Schaer.

On bark, etc. Barr Hill (Wilson).

* Var. **rosella,** Tuck.

Southwest Harbor (Eckfeldt).

* **L. tartarea** (L.), Ach.

On the earth. (Eckfeldt.)

* Var. **frigida,** Ach. Var. *telephoroides,* Th. Fries.

Thallus thin, whitish or cream-colored, papillose, ramulose becoming spinulose; sterile. On decayed grass and moss. (Eckfeldt.)

* **L. verrucosa** (Ach.), Laur.

On earth, etc. (Eckfeldt.)

L. cinerea (L.), Sommerf.

On rocks. Seal Harbor (Wilson).

* **L. molybdina** (Wahl.), Ach., var. **microcyclos,** Wahl.

On rocks. "Mt. Desert" (Tuckerman).

* **L. glaucocarpa** (Wahl.), Ach.

On rocks. (Eckfeldt.)

* **L. fuscata** (Schrad.), Th. Fries, var. **rufescens,** Th. Fries.

On rocks. (Eckfeldt.)

* **L. privigna** (Ach.), Nyl.

On rocks. (Eckfeldt.)

RINODINA, Mass.

* **R. oreina** (Ach.), Mass.

On rocks. (Eckfeldt, Wilson.)

* **R. sophodes** (Ach.), Nyl.

(Eckfeldt.)

* Var. **confragosa**, Nyl.
  On rocks. (Eckfeldt.)
* **R. constans** (Nyl.), Tuck.
  On old trees. (Eckfeldt.)

PERTUSARIA, DC.

P. **velata** (Turn.), Nyl.
On trees and rocks. (Eckfeldt); — Seal Harbor (Redfield).
* **P. leioplaca** (Ach.), Schaer.
  On trees and rocks. (Eckfeldt.)
* **P. pustulata** (Ach.), Nyl.
  On trees. (Eckfeldt.)

CONOTREMA, Tuck.

C. **urceolatum** (Ach.), Tuck.
On trees. Seal Harbor (Wilson).

GYALECTA, Ach.

* **G. lutea** (Dicks.), Tuck.
  On bark of trees. (Eckfeldt.)
* **G. Pineti** (Schrad.), Tuck.
  On bark of trees. (Eckfeldt.)

URCEOLARIA, Ach.

* **U. scruposa** (L.), Nyl.
  On rocks and earth. (Eckfeldt.)

THELOTREMA, Ach.

* **T. lepadinum**, Ach.
  On tree trunks. (Eckfeldt.)

TRIBE II. **LECIDEACEI.**

CLADONIEI.

STEREOCAULON, Schreb.

S. coralloides, Fries.
On rocks. (Eckfeldt); — Seal Harbor (Wilson).

S. paschale (L.), Fries.
On rocks. (Eckfeldt); — Seal Harbor (Wilson).

S. tomentosum (Fries), Th. Fries.
On ground. (Eckfeldt);—Seal Harbor (Wilson);—Southwest Harbor (M. L. Fernald).

* S. condensatum, Hoffm.
On earth. (Eckfeldt.)

* S. pileatum, Ach.
On rocks. (Eckfeldt.)

PILOPHORUS, Th. Fries.

* P. cereolus, Ach., var. Fibula, Tuck.
On rocks. (Eckfeldt.)

CLADONIA, Hoffm.

C. alcicornis (Lightf.), Floerke.
On the earth. Seal Harbor (Wilson).

* C. decorticata, Floerke.
On the earth. (Eckfeldt.)

C. pyxidata (L.), Fries.
On the earth. (Eckfeldt); — Seal Harbor (Redfield).

C. fimbriata (L.), Fries.
On the earth and rotten logs. (Eckfeldt, Wilson.)

Var. tubæformis, Fries.
On rotten logs. Seal Harbor (Redfield).

* Var. radiata, Fries.

On the earth.   Seal Harbor (Wilson).   ·

C. gracilis (L.), Nyl.

On the earth.   (Eckfeldt) ; — Seal Harbor (Wilson).

* Var. verticillata, Fries.

On the earth.   (Eckfeldt.)

Var. hybrida, Schaer.

On the earth.   Seal Harbor (Wilson).

Var. elongata, Fries.

On the earth.   (Eckfeldt); — Little Cranberry Isle (Redfield).
Forma macrocerus, Tuck., of this variety is also found (Eckfeldt).

C. turgida (Ehrh.), Hoffm.

On the earth.   (Eckfeldt); — Seal Harbor (Wilson).

C. papillaria (Ehrh.), Hoffm.

On the earth.   (Eckfeldt); — Barr Hill (Wilson).

C. cenotea (Ach.), Schaer.

On rotten logs and on the earth.   (Eckfeldt); — Seal Harbor (Wilson, Redfield).

C. squamosa, Hoffm.

On the earth, etc.   (Eckfeldt); — Seal Harbor (Wilson, Redfield).

C. cæspiticia (Pers.), Floerke.

On rocks and on the ground.   Seal Harbor (Wilson).

* C. furcata (Huds.), Fries.

On the earth.   (Eckfeldt, Wilson.)

Var. crispata, Floerke.

On the earth.   Seal Harbor (Wilson).

Var. racemosa, Floerke.

On the earth.   (Eckfeldt) ; — Seal Harbor (Wilson, Redfield).

* Var. subulata, Floerke.

On the earth.   (Wilson.)

C. rangiferina (L.), Hoffm.

Common on the earth, etc. (Eckfeldt, Wilson, Redfield.)

Var. **sylvatica,** L.

On rocks and earth. Seal Harbor (Wilson, Redfield).

Var. **alpestris,** L.

On the earth. (Wilson, Redfield.)

C. uncialis (L.), Fries.

On the earth. (Eckfeldt, Redfield.)

C. Boryi, Tuck.

On the earth. (Eckfeldt); — Seal Harbor (Wilson, Redfield).

C. coccifera (L.), Willd. *C. cornucopioides* (L.), Fries.

On the earth; common. (Eckfeldt, Wilson.)

* Var. **ochrocarpia,** Floerke.

(Eckfeldt.)

C. deformis (L.), Hoffm.

On the earth. (Eckfeldt, Redfield.)

C. digitata (L.), Hoffm.

On rotten wood and on the earth. (Eckfeldt, Wilson.)

C. macilenta (Ehrh.), Hoffm.

On the earth, etc. (Eckfeldt); — Seal Harbor (Redfield).

* **C. pulchella,** Schwein.

On old logs. (Eckfeldt.)

C. cristatella, Tuck.

On the earth, etc. (Eckfeldt); — Seal Harbor (Wilson, Redfield).

LECIDEEI.

BÆOMYCES, Pers.

* **B. byssoides** (L.), Schaer.

On clay soil. (Eckfeldt.)

B. roseus, Pers.

On ground; common. (Eckfeldt, Wilson, E. Faxon, Rand, Redfield.)

B. æruginosus (Scop.), DC.

On dead wood. (Eckfeldt, T. G. White.)

BIATORA, Fries.

* **B. rufo-nigra,** Tuck.

On rocks. (Eckfeldt.)

* **B. ostreata** (Hoffm.), Fries.

On carbonized pine wood. " Mt. Desert " (Willey).

* **B. coarctata** (Sm.), Nyl.

On rocks. (Eckfeldt.)

B. granulosa (Ehrh.), Poetsch.

On the earth, etc. Seal Harbor (Wilson).

* **B. parvifolia** (Pers.), Tuck.

On trees. (Eckfeldt.)

* **B. russula** (Ach.), Mont.

On tree trunks. (Eckfeldt.)

* **B. cinnabarina** (Sommerf.), Fries.

On tree trunks. (Eckfeldt.)

* **B. sanguineo-atra** (Fries), Tuck.

On the earth, etc. (Eckfeldt.)

* **B. varians,** Ach.

On bark and dead wood. (Eckfeldt.)

* **B. Laureri,** Hepp.

On trees. (Eckfeldt.)

* **B. hypnophila,** Turn.

On mosses, etc. (Eckfeldt.)

* **B. rubella** (Ehrh.), Rab.

On bark. (Eckfeldt.)

* **B. fusco-rubella,** Hoffm.

On trees.   (Eckfeldt.)

* **B. suffusa,** Fries.

On trees.   (Eckfeldt.)

HETEROTHECIUM, Flot.

H. sanguinarium (L.), Flot.

On tree trunks.   Northwest Arm woods (Rand).

H. pezizoideum (Ach.), Flot.

On bark.   Seal Harbor (Wilson).

LECIDEA, Ach.

* **L. contigua,** Fries.

On rocks.   (Eckfeldt.)

L. albocœrulescens (Wulf.), Schaer.

On rocks.   (Eckfeldt); — Seal Harbor (Wilson).

L. enteroleuca, Fries.

On rocks.   (Eckfeldt, Wilson.)

L. melancheima, Tuck.

On fence rails.   Seal Harbor (Wilson).

BUELLIA, De Not.

* **B. albo-atra** (Hoffm.), Th. Fries.

On trees and rocks.   (Eckfeldt.)

* **B. parasema** (Ach.), Th. Fries.

On trees and dead wood.   (Eckfeldt.)

* **B. myriocarpa** (DC.), Mudd.

On dead wood.   (Eckfeldt.)

* **B. colludens,** Nyl.

On rocks.   (Eckfeldt.)

B. petræa (Flot.), Tuck.

On rocks.   (Eckfeldt); — Seal Harbor (Wilson).

B. geographica (L.), Tuck.

On rocks.   Seal Harbor (Wilson).

TRIBE III.   **GRAPHIDACEI.**

OPEGRAPHEI.

OPEGRAPHA, Humb.*

* **O. demissa,** Tuck.

On smooth bark.   (Eckfeldt.)

* **O. viridis** (Pers.), Nyl.

Thallus light ochrous yellow, very thin, mostly scurfy or slightly tartareous, rimose, constricted.   Apothecia short, innate, from round to elongate and linear, mostly curved and undivided with the margin incurved.   Spores in thekes 8, hyaline, fusiform to acicular, 11 to 13-locular, $\frac{.035 - .056}{.007 - .008\frac{1}{2}}$ mic.   "Mt. Desert" (Eckfeldt).

XYLOGRAPHA, Fries.

* **X. disseminata,** Willey.

On dead wood.   "Mt. Desert" (Willey).

X. Opegraphella, Nyl.

On old fence rails.   Seal Harbor (Wilson).

GRAPHIS, Ach.

G. scripta (L.), Ach.

Thallus grayish white, very thin, membranaceous, tartareous even, and frequently rugose.   Apothecia slender, immersed, waving, margins narrow and elevated.   Spores hyaline, elongated, and cylindrical, 8–10-septate, $\frac{.0023 - .0025}{.006 - .007}$ mic.   Seal Harbor (Wilson, Eckfeldt).

---

* For description of genus, see Tuck., Syn. N. A. Lichens, part i., 11.

Var. **serpentina**, Ach.

Thallus ash-colored, tartareous, thicker than that of the preceding species, pulverulent, fissured, clearly determinate. Apothecia crowded, sunken, much elongated, simple or branched. Spores similar to preceding. On trees. Southwest Harbor (M. L. Fernald).

* **G. dendritica**, Ach.

Thallus cream-colored to a yellowish ash color, thin, tartareous, mealy. Apothecia brownish black, immersed, divided in branch-like clusters or pedate divisions. Spores long, linear, 6–8-septate, $\frac{.013-.015}{.001-.002}$ mic. Seal Harbor (Eckfeldt).

ARTHONIEI.

ARTHONIA, Ach.*

* **A. punctiformis**, Ach.

Thallus whitish, quite distinct or fading to a mere film. Apothecia very small, roundish or irregularly elongated, becoming variable in shape, often stellate. Spores long, ovoid, mostly 4-locular, $\frac{.015-.025}{.004-.008}$ mic. On tree barks. Southwest Harbor (Eckfeldt).

* **A. astroidea**, Ach.

Thallus pale ash-colored, opaque, finally almost wanting, but differing in color from the substrata, marginate with a waving terminal darkening line. Apothecia diverse, mostly stellate, slightly elevated, convex, pallid internally. Spores ovoid to elongated, 4-locular, $\frac{.008-.020}{.003-.007}$ mic. On tree barks. Southwest Harbor (Eckfeldt).

* Var. **Swartzoidea**, Nyl.

Thallus pale ash-colored, opaque but differing in color from the substrata, often becoming darker and more clearly distinct. Apothecia well pronounced, mostly round or irregular, $\frac{.010-.022}{.006-.010}$ mic. On tree barks. Southwest Harbor (Eckfeldt).

* For description of genus, see Tuck., Syn. N. A. Lichens, part i., 12.

* **A. Hamamelidis, Nyl.**

Thallus thin, white, but darkening, very diffused. Apothecia irregularly shaped, stellate, ramose, plane or somewhat convex, pale within. Spores 4-locular, $\frac{.009 - .012}{.003\frac{1}{2} - .004}$ mic. Generally on smooth barks of Hamamelis. Southwest Harbor (Eckfeldt).

TRIBE IV. CALICIACEI.

SPHÆROPHOREI.

SPHÆROPHORUS, Pers.* (*Sphærophoron*, Pers.)

S. globiferus (L.), DC. *S. coralloides*, Pers.

Thallus fruticulose, somewhat compressed or terete, with erect-ish, minutely fibrillose, ramulose branches, from pale to chestnut brown. Apothecia black, globose, shining, terminal with an inflexed margin. Spores spherical, violet black, medullary layer purple with potash reaction, .009 × .011 mic. "Mt. Desert" (Eckfeldt).

CALICIEI.

ACOLIUM, Fée.†

A. tigillare (Ach.), De Not.

Thallus yellow-green or more lemon-colored, granulose or areolate throughout. Apothecia black, stout, erect, more or less innate. Spores brown, ellipsoid to bilocular, $\frac{.0013 - .0023}{.007 - .011}$ mic. On old fence rails and boards. (Eckfeldt) ; — Seal Harbor (Wilson).

CALICIUM, Pers.‡

* **C. subtile, Fries.**

Thallus whitish or ash-colored, very thin, evanescent. Apothecia very small, stipe delicate, short and black, capitula very minute, somewhat globose, dark and depressed. Spores brown, 2-locular, $\frac{.006 - .009}{.002\frac{1}{2} - .003}$ mic. "Mt. Desert" (Eckfeldt).

---

* For description of genus, see Tuck., Syn. N. A. Lichens, part i., 13.

† For description of genus, see Ibid.

‡ For description of genus, see Ibid.

TRIBE V. **VERRUCARIACEI.**

ENDOCARPEI

ENDOCARPON, Hedw.*

* E. miniatum (L.), Schaer.

Thallus ashy white and darkening, simple, coriaceous, peltate, attached to the centre, the upper surface minutely granulose or pulvinate. Apothecia very minute, abundant, and enclosed in the thallus. Spores ellipsoid, simple, $\frac{.014 - .018}{.005\frac{1}{2} - .010}$ mic. On rocks; rather common. Southwest Harbor (Eckfeldt).

* Var. complicatum, Schaer.

Thallus closely cæspitosely conjoined, many growing together, border of thallus raised and frequently darkening, erect, complicate, often pruinose. Spores same as in type. On rocks. (Eckfeldt.)

VERRUCARIEI.

VERRUCARIA, Pers.†

* V. rupestris, Schrad.

Thallus grayish white or very pale, and even brown, very thin (becoming obsolete), tartareous and pulverulent. Apothecia black, numerous, sub-immersed but very conspicuous, hemispherical. Spores 8, hyaline, ellipsoid, oblong, simple, $\frac{.020 - .025}{.010 - .013}$ mic. On various rocks and stones. Southwest Harbor (Eckfeldt).

* V. nigrescens, Pers.

Thallus dark to a distinct black, rimulose, somewhat areolate, thick and uneven, loosely disposed and friable, developing into nodose elevations around the apothecia. Apothecia dull black, very numerous, hemispherical. Spores oblong, simple, $\frac{.018 - .024}{.007 - .009}$ mic. On rocks and stones. Southwest Harbor (Eckfeldt).

* For description of genus, see Tuck., Syn. N. A. Lichens, part i., 14.
† For description of genus, see Ibid., 15.

* **V. bryophila,** Lonnr.

Thallus whitish, encrusting certain mosses. Apothecia connate. Spores muriform. On the coast. Southwest Harbor (Eckfeldt).

PYRENULA, Ach.*

* **P. punctiformis** (Ach.), Naeg.

Thallus dark olivaceous, very thin, evanescent. Apothecia black, polished, very minute, conoid. Spores linear oblong, 2-locular, $\frac{.013-.015}{.003-.005}$ mic. On tree barks and dead wood. (Eckfeldt.)

* **P. nitida,** Ach.

Thallus pale yellow or olivaceous to brown, shining, smooth with scattered minute white nodules. Apothecia large, black. Spores $\frac{.019-.025}{.006-.008}$ mic. On smooth tree barks. Southwest Harbor (Eckfeldt).

P. thelæna, Ach.

Thallus uniform, mostly dark or fuscous. Apothecia small, innate. Spores colored, $\frac{.018-.023}{.007-.010}$ mic. Southwest Harbor (Eckfeldt).

* **P. lactea,** Mass.

Thallus cream-colored, very thin, spread out over the surface and limited. Apothecia black, minute, sessile, somewhat innate, becoming hemispherical. Spores broadly fusiform, 5–7-locular, $\frac{.0234-.0335}{.005-.007}$ mic. Seal Harbor (Eckfeldt).

* **P. gemmata** (Ach.), Naeg.

Thallus white, thin, continuous, smooth or rimulose. Apothecia black, large, prominent, roughish, hemispherical. Spores broadly oblong, 2-locular, $\frac{.013-.027}{.005\frac{1}{2}-.012}$ mic. On trees. Southwest Harbor (Eckfeldt).

---

* For description of genus, see Tuck., Syn. N. A. Lichens, part i., 15.

SUMMARY.

		Genera.	Species.	Varieties
Dicotyledones Angiospermeæ { Polypetalæ .		104	203	9
Gamopetalæ .		98	198	16
Apetalæ . .		24	59	5
Dicotyledones Gymnospermeæ		8	12	—
Monocotyledones		78	208	41
Pteridophyta		16	36	11
Bryophyta { Musci		41	166	56
Hepaticæ		26	48	1
Thallophyta { Characeæ		1	2	—
Algæ		86	140	6
Lichenes		45	214	59
Total		527	1286	204
Total of { Phanerogamia ; Flowering Plants		312	680	71
Cryptogamia; Flowerless Plants .		215	606	133

APPENDIX.

EXCLUDED SPECIES OF FLOWERING PLANTS, FERNS, AND FERN ALLIES.

Clematis verticillaris, DC. — Tunk Mt.,— on mainland (F. M. Day) ; beyond our limits.

Ranunculus BULBOSUS, L. — Northeast Harbor (Dunbar) ; doubtful.

Nymphæa odorata, Ait., var. minor, Sims. — Mountain Pond ; more correctly to be referred to the type.

Stellaria borealis, Bigel., var. alpestris (Fries), Gray. — Bear Island (Redfield) ; = type.

Cerastium nutans, Raf. — (F. M. Day). The specimen is *C. vulgatum*, L.

Hypericum corymbosum, Muhl. — Roadside between Somesville and Southwest Harbor (Dunbar) ; never verified.

Malva VERTICILLATA, L. — Waste ground, Long Pond (Rand) ; not persistent, only casual.

Tilia INTERMEDIA, Hayne. — Somesville (Redfield) ; only in cultivation.

Linum HUMILE, Mill. — Southwest Harbor (Rand). The specimen is *L. usitatissimum*, L.

Geranium maculatum, L. — Several times reported, but never verified. Probably confused with *G. Carolinianum*, L.

Geum Virginianum, L. — In fruit; Somesville (Rand). Flowering specimens from same station show this to be *G. strictum*, Ait.

Cratægus tomentosa, L. — Reported by various collectors, but is undoubtedly *C. coccinea*, L. var. *macracantha* (Lodd.), Dudley.

Ribes NIGRUM, L. — Somesville (R. & R.). The specimens are *R. floridum*, L'Hér.

Epilobium alpinum, Gray, Man., 5th ed. — Newport Mt. (F. M. Day) ; never verified.

Angelica atropurpurea, L.

Little Cranberry Isle (Wakefield); never verified.

Cryptotænia Canadensis, DC.

Southwest Harbor (Wakefield); never verified.

Conium MACULATUM, L.

Never verified.

Solidago humilis, Pursh., var. microcephala, Porter.

Frenchman Camp road (Redfield); only a pecular form of *S. nemoralis*, Ait. (Bull. Torr. Bot. Club., xx. 210.)

Solidago puberula, Nutt., var. monticola, Porter.

This is *S. Virgaurea*, L., var. *monticola*, Porter. (Bull. Torr. Bot. Club, xx. 209.)

Solidago speciosa, Nutt.

Roadside between Seal Cove and Norwood Cove (Wakefield); never verified. Probably *S. Virgaurea*, L., var. *Randii*, Porter.

Solidago odora, Ait.

Several times reported, but never verified.

Solidago juncea, Ait., var. ramosa, Porter & Britt.

This is *S. puberula*, Nutt., *forma.*

Solidago rupestris, Raf.

This is *S. Canadensis*, L., var. *glabrata*, Porter.

Callistephus CHINENSIS, DC.

A garden escape, Southwest Harbor; not persistent.

Aster Herveyi, Gray.

Near Bubble Pond (Redfield). This is undoubtedly a small form of *A. macrophyllus*, L.

Erigeron annuus (L.), Pers.

Roadside near Fernald Point (Wakefield); never verified.

Senecio CINERARIA, DC.

Shore, Mt. Desert Narrows (R. & R.). The specimens are *Artemisia Stelleriana*, Bess.

Cnicus pumilus (Nutt.), Torr.

Cranberry Isles (Harriet A. Hill); never verified.

Lampsana COMMUNIS, L.

Sutton Island (Wakefield); never verified.

Prenanthes alba, L.

Often reported, but never verified All specimens examined are *P. serpentaria*, Pursh.

Vaccinium vacillans, Solander.

Northeast Harbor; Hulls Cove (Curtis). This is doubtless *V. Canadense*, Kalm.

Anagallis ARVENSIS, L.

Hulls Cove (Annie S. Downs); never verified.

Fraxinus pubescens, Lam.

Never verified.

Convolvulus ARVENSIS, L.

Schooner Head (F. M. Day); never verified.

Limosella aquatica, L., var. tenuifolia (Nutt.), Hoffm.
Shore, Mt. Desert Narrows (F. M. Day); never verified.

Veronica LONGIFOLIA, L.
Sea Wall road, Southwest Harbor (Rand). A garden escape; not persistent.

Euphrasia officinalis, L., var. Tatarica, Benth.
Sea Wall (Rand). This is some other form of the protean type.

Utricularia biflora, Lam.
Somes Pond (Rand). The specimen is *U. gibba*, L. See note *ante*, under that species.

Mentha Canadensis, L., var. glabrata, Benth.
Somesville (Rand). The specimens are *M. arvensis*, L., *forma*.

Plantago major, L., var. minima, (DC.), Decsne.
Near Little Harbor; Great Head (Redfield). Specimens prove to be some other form of the type.

Atriplex arenaria, Nutt.
Sea Wall (H. C. Jones); — Somes Sound (Redfield); never verified.

Polygonum erectum, L.
Beach, Wasgatt Cove (Lane); never verified, — doubtless *P. Raii*, Bab.

Polygonum maritimum, L.
Reported from various stations (Elizabeth G. Britton, and others). Specimens are *P. Raii*, Bab.

Polygonum hydropiperoides, Mx.
Little Cranberry Isle (Rand); — Seal Harbor (Redfield); too doubtful, — never verified.

Polygonum arifolium, L.
Bar Harbor (F. M. Day); too doubtful, — never verified.

Polygonum dumetorum, L., var. scandens, Gray.
Reported by various collectors, but all specimens are *P. Convolvulus*, L.

Humulus Lupulus, L.
Seal Harbor (Redfield); — Bar Harbor (W. H. Manning); introduced; hardly an escape.

Urtica DIOICA, L.
Sea Wall (H. C. Jones); never verified, — doubtless *U. gracilis*, Ait.

Alnus serrulata, Willd.
Near Hadlock Upper Pond (Rand); too doubtful, — never verified.

Quercus coccinea, Wang.
All specimens are *Q. rubra*, L.

Pinus Banksiana, Lamb.
Only on and about Schoodic Peninsula (F. M. Day, Rand, Redfield); beyond our limits.

Juniperus communis, L., var. alpina, Gaud.
Mt. Desert forms so far as seen are rather to be referred to the type.

Liparis liliifolia (L.), Richard.
Accredited to Mt. Desert in Baldwin's "Orchids of New England," but never verified.

Aplectrum hiemale, Nutt.
Reported from near Bar Harbor, 1882 (F. M. Day), but never verified.

Spiranthes latifolia, Torr. — Near Bass Harbor (W. H. Dunbar); never verified, — doubtless *S. Romanzoffiana*, Cham.

Orchis spectabilis, L. — Southwest Harbor (H. M. Pratt); never verified.

Habenaria psycodes (L.), Gray. — No station (W. H. Dunbar); never verified, but doubtless a small form of *H. fimbriata* (Ait.), R. Br.

Iris prismatica, Pursh. — Great Cranberry Isle (W. H. Dunbar); never verified.

Sisyrinchium anceps, Cav. — "Mt. Desert" (F. M. Day). The specimen is apparently an immature plant of *S. angustifolium*, Mill.

Uvularia perfoliata, L. — Salisbury Cove (Rand); never verified, — doubtless an error.

Trillium erectum, L. — Green Mt. Gorge (F. M. Day); never verified.

Xyris flexuosa, Muhl. — Hadlock Lower Pond (W. H. Dunbar); never verified, — doubtless only var. *pusilla*, Gray, which has been found at the same station.

Xyris Caroliniana, Walt. — Breakneck Ponds (R. H. Day); = *X. flexuosa*, Muhl., var. *pusilla*, Gray.

Luzula spicata, Desv. — Northeast Harbor (Greenleaf); never verified.

Carex pubescens, Muhl. — "Mt. Desert" (F. M. Day). The specimen is *C. communis*, Bailey.

Carex adusta, Boott, var. glomerata, Bailey. — Northeast Harbor (Greenleaf). This is *C. fœnea*, Willd.

Muhlenbergia sylvatica, T. & G. — Northeast Harbor (Greenleaf). The specimen is *M. glomerata*, Trin.

Avena SATIVA, L. — Occasionally appearing; not persistent.

Poa alsodes, Gray. — Little Cranberry Isle (Lane); — Northeast Harbor (Greenleaf). The specimens are *P. serotina*, Ehrh.

Puccinellia distans (Wahl.), Parl. — Northeast Harbor (Greenleaf); = *P. maritima* (Wahl.), Parl., var. (?) *minor*, Wats.

Agropyrum dasystachyum (Gray), Vasey. — No station (Greenleaf); never verified. Doubtless a form of *A. repens* (L.), Beauv.

Adiantum pedatum, L. — Often reported, but never verified.

Woodwardia Virginica, Smith. — Banks of Hadlock Brook (Wakefield); never verified.

Cystopteris bulbifera, Bernh. — Sargent Mt. Gorge (Rand); an error.

INDEX.

GENERA AND COMMON NAMES.

[SYNONYMS IN ITALICS.]

Abies	149	Arbor-Vitæ	150	Bearberry	125
Acer	89	Arbutus, Trailing	126	Bedstraw	108
Achillea	118	Arctium	120	Beech	145
Acolium	272	Arctostaphylos	125	Beechdrops	134
Acorus	160	Arenaria	83	Beggar Ticks	118
Actæa	76	Arethusa	152	Bellflower	124
Adder's Mouth	150	Arisæma	159	Bellwort	156
Adder's Tongue	188	Arnoseris	120	Berberis	77
Adiantum	280	Arrowhead	160	Betula	144
Agarum	236	Arrow-wood	107	Biatora	268
Agrimonia	97	Artemisia	119	Bidens	118
Agrimony	97	Arthonia	271	Bindweed	131
Agropyrum	182, 280	Arum, Water	160	Black	142
Agrostis	177	Ascocyclus	240	Birch	144
Ahnfeldtia	232	Ascophyllum	235	Bittersweet	131
Alaria	236	Ash	129	Blackberry	94
Alder	144	Mountain	98	Black-eyed Susan	117
Black	88	Asparagus	155	Bladderwort	134
White	127	Aspen	147	Blepharostoma	221
Alectoria	253	Asperococcus	239	Blinks	85
Alnus	144, 279	Aspidium	185	Blite, Sea	140
Alopecurus	177	Asplenium	184	Bluebell	124
Alsike	91	Aster	113, 278	Blueberry	124
Amaranth	139	Astrophyllum	208	Bluebottle	120
Amarantus	139	Atrichum	209	Blue Joint	178
Ambrosia	117	Atriplex	140, 279	Bluets	108
Amelanchier	99	Atropis	181	Boneset	109
Ampelopsis	89	Aulacomnium	209	Botrychium	187
Amphicarpæa	93	Avena	280	Brachyelytrum	176
Amphoridium	205	Avens	95	Brake	184
Anagallis	278	Azalea	126	Brasenia	77
Anaphalis	117			Brassica	79
Andreæa	199	Bachelor's Button	120	Brier, Sweet	97
Andromeda	126	Bæomyces	267	Bromus	182
Anemone	75	Baked Apple Berry	94	Broom-rape	134
Star	128	Balm of Gilead	148	Brunella	137
Aneura	225	Balsam	88	Bryum	207
Angelica	278	Balsam Apple, Wild	103	Buckbean	130
Antennaria	117	Baneberry	76	Buckwheat	142
Anthemis	118	Bangia	234	Buda	84
Anthoxanthum	176	Barbarea	79	Buellia	269
Antithamnion	230	Barberry	77	Bugleweed	136
Aphyllon	134	Barbula	204	Bugloss	130
Aplectrum	279	Barley	183	Bulbocoleon	244
Apocynum	129	Bartonia	130	Bunchberry	106
Apple	98	Bartramia	207	Burdock	120
Apple of Peru	131	Batrachospermum	233	Bur Marigold	118
Aquilegia	76	Bayberry	143	Bur-reed	159
Aralia	106	Bazzania	220	Butter-and-Eggs	132

Buttercup	75	Cleavers	109	Dicksonia	186
Butterweed	116	Clematis	75, 277	Dicranella	200
		Clethra	127	Dicranum	201
Cabbage, Skunk	160	Climacium	213	Dictyosiphon	239
Cakile	80	Clintonia	156	Diervilla	108
Calamagrostis	178	Clover	90	Diphyscium	211
Calicium	272	Club-moss	188	Diplophyllum	222
Calla	160	Cnicus	120, 278	*Ditrichum*	204
Calla Lily, Wild	160	Cockle	82	Dock	140
Callistephus	278	Codiolum	242	Dodder	131
Callithamnion	230	Cœlopleurum	105	Dogbane	129
Callitriche	101	Collema	260	Dogwood	106
Calopogon	152	Coltsfoot, Sweet	119	Doorweed	141
Calothrix	246	Columbine	76	Drosera	100
Campanula	124	*Comptonia*	143	Dulichium	162
Campion	82	Cone Flower	117	Dulse	232
White	82	Conferva	244	Dusty Miller, False	119
Capsella	79	Conioselinum	104		
Capsosiphon	245	Conium	278	Echinocystis	103
Caraway	105	Conocephalus	226	Ectocarpus	240
Cardamine	78	Conotrema	264	Elachistea	238
Cardinal Flower	123	Convolvulus	131, 278	Elatine	86
Carex	165, 280	Coptis	76	Elder	107
Carpet Weed	104	Coral-root	151	Eleocharis	162
Carrot	104	Corallina	228	Elm	143
Carum	105	Corallorhiza	151	Elodes	87
Cassandra	126	Corema	148	Elymus	183
Castagnea	238	Cornel	106	Empetrum	148
Catchfly	82	Cornus	· 106	Endocarpon	273
Catharinea	209	Corydalis	77	Enteromorpha	245
Catnip	137	Corylus	144	*Entodon*	212
Cedar, White	150	Cow Herb	82	Epigæa	126
Centaurea	120	Cow-lily	77	Epilobium	102, 277
Cephalozia	221	Cranberry	124	Epiphegus	134
Ceramium	230	Mountain	125	Equisetum	184
Cerastium	84, 277	Cratægus	98, 277	Erechtites	120
Ceratodon	203	Cress, Bitter	78	Erigeron	116, 278
Cetraria	251	Marsh	79	Eriocaulon	162
Chætomorpha	243	Water	79	Eriophorum	164
Chamomile	118	Winter	79	Erythronium	156
Chantransia	233	Crowberry	148	Eupatorium	109
Charlock	79	Broom	148	Euphorbia	142
Jointed	80	Crowfoot	75	Euphrasia	133, 279
Checkerberry	126	Cryptotænia	278	Euthora	232
Cheeses	87	Cucumber-root, Indian	156	Everlasting	117
Chelone	132	Cudweed	117	Pearly	117
Chenopodium	140	Currant	99	Sweet	117
Cherry	93	Cuscuta	131	Evernia	252
Chickweed	83	Cynodontium	200	Eyebright	133
Field	84	Cypripedium	153		
Indian	104	Cystoclonium	232	Fagopyrum	142
Mouse-ear	84	Cystopteris	186, 280	Fagus	145
Chicory	121			Fern, Beech	185
Chiloscyphus	223	Dactylis	179	Bladder	186
Chimaphila	127	Daisy, Ox-eye	119	Christmas	186
Chiogenes	125	Yellow	117	Cinnamon	187
Chokeberry	98	Dalibarda	95	Flowering	187
Chondrus	232	Dandelion	122	Lady	184
Chorda	237	Fall	121	Sensitive	186
Chordaria	237	Danthonia	179	Shield	185
Choreocolax	233	Daucus	104	Sweet	143
Christmas Green, Trailing	189	Decodon	102	Festuca	181
Chroococcus	249	Delesseria	231	Fir	149
Chrysanthemum	119	Dermocarpa	249	Fireweed	102, 120
Chrysosplenium	99	Deschampsia	178	Fissidens	203
Cichorium	121	Desmarestia	238	Five-finger	96
Cicuta	105	Desmodium	92	Flag, Blue	154
Cinna	178	Desmotrichum	239	Cat-tail	159
Cinquefoil	96	Devil's Apron	236	Sweet	160
Circæa	103	Devil's Pitchfork	118	Flax	87
Cladium	165	Dewberry	95	Toad	132
Cladonia	265	Dianthus	82	Fleabane	116
Cladophora	242	Dichelyma	211	Daisy	116

Floating Heart 130
Fontinalis 211
Fossombronia 225
Fragaria 96
Fraxinus 129, 278
Frullania 220
Fucus 235
Fumaria 78
Fumitory 78
Funaria 206

Gale, Sweet 143
Galeopsis 138
Galium 108
Gaultheria 126
Gaylussacia 124
Geocalyx 223
Georgia 206
Geranium 88, 277
Germander 135
Geum 95, 277
Gigartina 232
Gill-over-the-Ground 137
Glaux 129
Gloeocapsa 249
Gloeosiphonia 229
Glyceria 180
Gnaphallum 117
Gold Buttons 119
Golden Rod 109
Goldthread 76
Gomontia 242
Goodyera 152
Gooseberry 99
Graphis 270
Grass, Arrow 161
Barn-yard 175
Bent 177
Black 158
Blue-eyed 154
Bristly Foxtail 175
Brome 182
Cat's-tail 177
Cotton 164
Cut 175
Ditch 162
Drop-seed 176, 177
Eel 162
Feather 176
Fescue 181
Foxtail 175, 177
Hair 178
Herd's 177
Kentucky Blue 179
Lyme 183
Manna 180
Marsh 174
Meadow 179
Mist 177
Orange 87
Orchard 179
Panic 174
Quitch 182
Rattlesnake 180
Reed Bent 178
Salt Rush 174
Sea Spear 181
Spear 179
Squirrel-tail 183
Sweet 176
Sweet Vernal 176
Wild Oat 179
Wire 179

Grass, Witch 182
Wood Reed 178
Yellow-eyed 157
Graveyard Flower 142
Grimmia 204
Groundsel 119
Gyalecta 264

Habenaria 153, 280
Hackmatack 149
Halosaccion 235
Hamamelis 101
Hardhack 94
Harebell 124
Hawkweed 121
Hawthorn 98
Hazel-nut 144
Hazel, Witch 101
Heart's-ease 81
Hedeoma 137
Hedwigia 205
Helianthus 118
Hemerocallis 154
Hemlock 149
Ground 150
Water 105
Heppia 259
Heracleum 104
Herb Robert 88
Heterothecium 269
Hieracium 121
Hierochloe 176
Hildenbrandtia 235
Hippuris 101
Hobble Bush 107
Holly 88
Mountain 89
Honeysuckle 108
Bush 108
Fly 108
Hop Clover 91
Hordeum 183
Horehound, Water 136
Horseradish 79
Horsetail 184
Horseweed 116
Houstonia 108
Huckleberry 124
Hudsonia 80
Humulus 279
Hydrocotyle 105
Hyella 249
Hypericum 86, 277
Hypnum 213

Ilex 88
Ilysanthes 132
Impatiens 88
Indian Cucumber-root 156
Indian Pipe 128
Innocents 108
Iris 154, 280
Isoetes 189
Ivy, Ground 137
Poison 90

Jack-in-the-Pulpit 159
Jewel Weed 88
Job's Tears 180
Joe-Pye Weed 109
Jubula 220
Juncus 157
Jungermannia 223

Juniper 150
Juniperus 150, 279

Kalmia 126
Kantia 222
Knotweed 141

Labrador Tea 127
Lactuca 122
Ladies' Delight 81
Ladies' Tresses 151
Lady's Slipper 153
Lady's Thumb 142
Lambkill 126
Laminaria 236
Lampsana 278
Larch 149
Larix 149
Lathyrus 93
Laurel, American 126
Pale 126
Sheep 126
Lavender, Sea 128
Leather-leaf 126
Leathesia 238
Lecanora 262
Lechea 80
Lecidea 269
Ledum 127
Leersia 175
Lemanea 234
Leontodon 121
Leonurus 138
Lepidium 80
Lepidozia 221
Leptobryum 207
Leptogium 261
Leptotrichum 204
Leskea 212
Lettuce 122
Sea 246
Leucobryum 203
Leucodon 212
Ligusticum 105
Lilac 129
Lilium 156
Lily 156
Day 154
Lily of the Valley, Wild 156
Limnanthemum 130
Limosella 279
Linaria 132
Linnaea 107
Linum 87, 277
Liparis 150, 279
Listera 151
Lithophyllum 228
Lithothamnion 228
Live-forever 100
Liverworts 219, 226
Lobelia 123
Locust-tree 92
Lonicera 108
Loosestrife 128
False 102
Swamp 102
Lophocolea 223
Lousewort 134
Lovage 105
Ludwigia 102
Lungwort 130
Sea 130
Luzula 158, 280

Lychnis	82	Neckera	212	Picea	149
Lycopodium	188	Nemalion	233	Pickerel-weed	157
Lycopsis	130	Nemopanthes	89	Pigweed	140
Lycopus	136	Nepeta	137	Amaranth	137
Lyngbya	247	Nephroma	258	Pilophorus	265
Lysimachia	128	Nettle	143	Pimpernel, False	132
		Hedge	138	Pine	149
Madotheca	220	Hemp	138	Ground	188
Maianthemum	155	Nicandra	131	Pinesap	128
Mallow	87	Nightshade	131	Pine Weed	87
Malva	87, 277	Enchanter's	103	Pink	82
Maple	89	Nitella	227	Swamp	152
Marchantia	226	Nuphar	77	Pinus	149, 279
Mare's Tail	101	Nymphæa	77, 277	Pinweed	80
Marsupella	225			Pipe, Indian	128
Mastigobryum	220	Oak	144	Pipewort	162
Mastigocoleus	247	Oakesia	156	Pipsissewa	127
Mayflower	126	Œnothera	103	Pitcher-plant	77
Mayweed	118	Omphalaria	260	Placodium	261
Meadow Beauty	102	*Oncophorus*	200	Plagiochila	223
Meadow Rue	75	Onoclea	186	Plantago	138, 279
Meadow Sweet	94	Opegrapha	270	Plantain	138
Medeola	156	Ophioglossum	188	Rattlesnake	152
Medicago	92	Orache	140	Platygyrium	212
Medick	92	Orchis	280	Plumaria	230
Melampyrum	134	Purple-fringed	153	Poa	179, 280
Melilotus	91	Orpine	100	Pogonatum	209
Melilot	91	Orthotrichum	206	Pogonia	152
Melobesia	229	Oryzopsis	176	*Pohlia*	207
Mentha	136, 279	Oscillatoria	248	Polycystis	249
Menyanthes	130	Osmunda	187	Polygala	90
Mermaid Weed	101	Oxalis	88	Polygonatum	155
Mertensia	130			Polygonum	141, 279
Microcoleus	247			Polyides	229
Microstylis	150	Pallavicinia	225	Polypodium	184
Milfoil, Water	101	*Panicularia*	180	Polypody	184
Milkwort	90	Panicum	174	Polysiphonia	231
Sea	129	Pannaria	260	Polytrichum	210
Millet	175	Pansy	81	Pond-lily, Yellow	77
Mint	136	Parietaria	143	Pondweed	161
Mitchella	108	Parmelia	254	Pontederia	157
Mitella	99	Parsley, Hemlock	104	Poplar	147
Mitrewort	99	Parsnip	104	Populus	147
Mnium	208	Cow	104	Porella	220
Mollugo	104	Water	105	Porphyra	234
Moneses	127	Partridge Berry	108	Portulaca	85
Monostroma	246	Pastinaca	104	Potamogeton	161
Monotropa	128	Pea, Beach	93	Potentilla	96
Montia	85	Everlasting	93	Prenanthes	122, 278
Moonwort	187	Field	93	Primrose, Evening	103
Morning Glory, Wild	131	Marsh	93	Prince's Pine	127
Moss, Club	188	Peanut, Hog	93	Proserpinaca	101
Irish	232	Pearlwort	84	Protococcus	241
Peat	191	Pear, Sugar	99	Prunus	93
Mosses, Scale	220	Pedicularis	134	Pteris	184
Motherwort	138	Pellia	225	Ptilidium	220
Mouse-ears	117	Pellitory	143	Ptilota	230
Mugwort	119	Peltigera	259	Puccinellia	181, 280
Muhlenbergia	176, 280	Pennyroyal, American	137	Punctaria	239
Mullein	131	Pennywort, Water	105	Purslane	85
Mustard, Black	79	Pepperbush, Sweet	127	Water	102
Hedge	79	Peppergrass	80	Pursley	85
Mylia	223	Pepper, Water	142	Pylaiella	241
Myrica	143	Pertusaria	264	Pylaisia	212
Myrionema	238	Petasites	119	Pyrenula	274
Myriophyllum	101	Petrocelis	229	Pyrola	127
Myurella	212	Peyssonnelia	229	One-flowered	127
		Phegopteris	185	Pyrus	98
Nabalus	122	Philonotis	207	Pyxine	257
Naiad	162	Phleum	177		
Naias	162	Phormidium	248	Quaker Ladies	108
Nardia	225	Phyllitis	239	Quercus	144, 279
Nasturtium	79	Physcia	256	Quillwort	189

Racomitrium	204	Scutellaria	137	Sticta	258
Radish	80	Scytosiphon	239	Stigonema	247
Ragweed	117	Sedge	165	Stipa	176
Ragwort, Golden	119	Sedum	100	St. John's-wort	86
Ralfsia	237	Selaginella	189	Marsh	87
Ramalina	250	Self-heal	137	Stonecrop	100
Ranunculus	75, 277	Senecio	119, 278	Strawberry	96
Raphanus	80	Setaria	175	Streptopus	155
Raspberry	94	Shadbush	99	Suæda	140
Rattlesnake-root	122	Shepherd's Purse	79	Succory, Lamb's	120
Rattle, Yellow	133	Shinleaf	127	Sumach	90
Red-top	178	Side-saddle Flower	77	Sundew	100
False	179	Silene	82	Sunflower	118
Rhexia	102	Silver Weed	96	Sweet Brier	97
Rhinanthus	133	Sisymbrium	79	Sweet Gale	143
Rhizoclonium	243	Sisyrinchium	154, 280	Symplocarpus	160
Rhodochorton	230	Sium	105	Syringa	129
Rhododendron	126	Skullcap	137		
Rhodomela	231	Smilacina	155	Tamarack	149
Rhodora	126	Snails	92	Tanacetum	119
Rhodymenia	232	Snake-head	132	Tansy	119
Rhus	90	Snakeroot, Black	106	Taraxacum	122
Rhynchospora	165	Sneezewort	118	Tare	92
Ribes	99, 277	Snowberry, Creeping	125	Taxus	150
Ribgrass	138	Solanum	131	Tear-thumb	142
Rice, Mountain	176	Solidago	109, 278	Tetranema	244
Rinodina	263	Solomon's Seal	155	Tetraphis	206
Rivularia	247	Dwarf	155	Teucrium	135
Robinia	92	False	155	Thalictrum	75
Rocket, Sea	80	Solorina	259	Theloschistes	254
Yellow	79	Sonchus	123	Thelotrema	264
Rockweed	235	Sorrel	140	Thistle	120
Rosa	97	Field	141	Sow	123
Rose	97	Wood	88	Star	120
Rose-bay	126	Sparganium	159	Thorn	98
Rosemary, Marsh	128	Spartina	174	Thoroughwort	109
Roseroot	100	Spatter Dock	77	Thuja	150
Rubus	94	Spearmint	136	Thyme	136
Rudbeckia	117	Spearwort, Creeping	76	Thymus	136
Rumex	140	Specularia	123	Tick Trefoil	92
Ruppia	162	Speedwell	132	Tilia	277
Rush	157	Spergula	85	Timothy	177
Beak	165	Spergularia	84	Toad Flax	132
Club	163	Sphacelaria	240	Tobacco, Indian	123
Scouring	184	Sphænosiphon	249	Ladies'	117
Spike	162	Sphærocephalus	209	Touch-me-not, Spotted	88
Twig	165	Sphærophoron	272	Trefoil	90
Wood	158	Sphærophorus	272	Tick	92
Rye, Wild	183	Sphagnum	191	Trematodon	200
		Spikenard	106	Trentepohlia	233
Saccorhiza	236	False	155	Trichocolea	220
Sage, Wood	135	Spinach	139	Trientalis	128
Sagina	84	Spinacia	139	Trifolium	90
Sagittaria	160	Spiræa	94	Triglochin	161
Salicornia	140	Spiranthes	151, 280	Trillium	156, 280
Salix	145	Spirogyra	241	Trisetum	179
Salsola	140	Spirulina	248	Triticum	182
Saltwort	140	Splachnum	206	Tsuga	149
Sambucus	107	Spleenwort	184	Tumbleweed	139
Samphire	140	Sporobolus	177	Tuomeya	234
Sandwort	83	Spruce	149	Turnip	79
Sanicula	106	Spurge	142	Indian	159
Saponaria	82	Spurrey	85	Turtle-head	132
Sarracenia	77	Corn	85	Tway-blade	150, 151
Sarsaparilla, Wild	106	Sand	84	Twin Flower	107
Satureia	136	Stachys	138	Twisted Stalk	155
Savory	136	Star-flower	128	Typha	159
Saxifraga	99	Starwort	83		
Saxifrage	99	Water	101	Ulmus	143
Golden	99	Statice	128	Ulota	205
Scapania	222	Steetzia	225	Ulothrix	244
Scheuchzeria	161	Stellaria	83, 277	Ulva	246
Scirpus	163	Stereocaulon	265	Umbilicaria	257

Urceolaria	264	Virginian Creeper	89	Wintergreen, Creeping	126
Urtica	143, 279	Virgin's Bower	75	Witch Hazel	101
Usnea	252			Withe-rod	107
Utricularia	134, 279	Wake Robin	156	Woodbine	89
Uvularia	156, 280	Water-lily	77	Woodsia	186
		Water Shield	77	Woodwardia	280
Vaccinium	124, 278	Waterwort	86	Wormwood	119
Vaucheria	241	Webera	207	Woundwort	138
Venus's Looking-glass	123	Webera	211		
Verbascum	131	Weissia	205	Xylographa	270
Veronica	132, 279	Wheat, Cow	134	Xyris	157, 280
Verrucaria	273	False	182		
Vetch	92	White-weed	119		
Viburnum	107	Willow	145	Yarrow	118
Vicia	92	Willow Herb	102	Yellow Rattle	133
Viola	80	Wind-flower	75	Yew	150
Violet	80	Wintergreen	127		
Dog-tooth	156	Aromatic	126	Zostera	162